Jerry Jordan
11357 Yale Bridge R
Amboy, WA 98601

D1792474

THE HYDROGEN WORLD VIEW

First Edition

Dr. Roger E. Billings

AMERICAN ACADEMY OF SCIENCE
Independence, Missouri

Copyright 1991 by
American Academy of Science
TechCenter, Suite 1000
26900 East Pink Hill Road
Independence, Missouri 64057-3284

Library of Congress Catalog Card Number: 91-77512

ISBN: 0-9631634-0-x

All rights reserved. No part of this book may be
reproduced, stored in a retrieval system, or
transcribed in any form or by any means, electronic
or mechanical, including photocopying and recording,
without the prior written permission of the publisher.

Printed in the United States of America

THE HYDROGEN WORLD VIEW

First Edition

DEDICATION

This book is dedicated to my lovely wife, Tonja A. Billings, who not only lived through the experiments outlined in this volume, but also typed the original manuscript. Thank you Tonja!

I would like to express my special thanks to the following individuals for their valuable and dedicated assistance in preparing this work:

    Dr. Maria Sanchez - Technical critique and illustrations
    Eileen Dayton - Editing and "creative" English
    Jan Eyre and Marci Merkley - Graphics and layout

FOREWORD

In 1962, Rachel Carson expressed her disappointment that bird song had almost ceased in a famous book, Silent Spring. In 1969, research results were published which tied the smog in Los Angeles unambiguously to the emissions from automotive exhausts. In the 1970's the news media and the public began to get involved in and worried about the environment and the automobile. Finally, in the 80's, the realization has spread--the automobile, driven by gasoline, is making the entire planet warm up.

For much of this time--from the early 70's--the author of this book, first as a boy in his back room experiments, then in a garage, then in his own business, and finally, now, as a successful head of several corporations, has led the race to make practical the use of clean hydrogen as a medium of energy in transportation.

His career and my own interest in hydrogen--that, too, of my close colleague, T. Nejat Veziroglu--have overlapped in time. But one of the differences between Roger Billings on the one side and Veziroglu and me on the other, is that Billings has always been the one who did things. He seldom talked about them. Veziroglu and I have (from '74 until now) researched, organized meetings, and propagandized the concept of the Hydrogen Economy in lectures throughout the world. All this time Billings was working, working on the weekends and working late at night, particularly to realize the goal which he had had in mind since his youthhood, the Billings actualization of the concept of a Hydrogen Economy.

At first there was the home in Utah, completely a hydrogen home in all ways, and then the move to Independence, Missouri, and the formation of corporate entities from which so much has come--and from which, one hopes, the major part is still to come.

I am delighted that Roger Billings has at last had time to sit down and tell his story. The book is a splendidly informal mixture of a description of scientific and engineering accomplishments, but also the human story, the booms and busts of the researcher's life in a career set concentratedly upon the one goal--to bring a clean fuel to the public in an economically acceptable condition.

Billings has clearly reached a new height with his car which can be charged with hydrogen from a normal household electricity outlet, and which reconverts to electricity with the efficiency which can only be obtained from fuel cells, and which offers clean motoring at a net cost not exceeding that of the present polluting system.

What a triumph that this individual man has been able to do what has escaped the ability of the world's large corporations--to actualize an effective contribution to the most important area of all, the economically acceptable application of clean hydrogen to automotive transportation.

Roger Billings is still young. There are many more years for him to achieve in and give more, and all of it will be in work carried out in an ordered and determined way with a decisive program and a relentless marching to the goal.

> J. O'M. Bockris
> Distinguished Professor of Chemistry
> Texas A&M University

PREFACE

This book is written in response to the requests of literally thousands of hydrogen enthusiasts. Having become a full-time hydrogen energy devotee at the age of 14, my role in the emerging hydrogen energy economy seems to have become the practical demonstration and implementation of technologies which so many people had previously declared to be impossible. My purpose in writing this book is to provide, in a simple-to-follow, hands-on, first person format, a history of my 25-year dedication to this work. I plan to include personal experiences (sometimes irrelevant), experiments in sufficient detail to enable the enthusiastic reader to recreate and improve upon such, and an account of the evolution of the technology, along with my personal opinions concerning that technology.

It is my goal to make available to my readers as much information as possible, thereby enabling new hydrogen supporters to get their feet wet in this new technology in as short a time as possible. If successful, I expect this book to be useful to a wide range of readers, from high school students working on hydrogen-related science fair projects, to PhD's having already completed their formal education. Although no attempt is made to document historical facts, (I have no pretense of being an historian), a lot of historical information will be provided in an attempt to help the reader understand the research and the reasons for specific experiments.

Although the evolution of the hydrogen energy economy is the result of the combined efforts of hundreds of capable scientists, I make no attempt to acknowledge or document the contributions of my wonderful colleagues. I trust they will realize that this in no way diminishes my opinion of their work or constitutes a failure of recognition on my part of the important contributions each has made. In fact, I sincerely encourage them to follow my example by writing their own histories, an autographed copy of which I look forward to expectantly.

Well then, this is my story. I will tell it as I have lived it--simply, sometimes arrogantly, and always with enthusiasm. If even one reader catches the dream and joins the Hydrogen Energy Team, then our race with time to keep this planet a nice place to live will benefit. It is to this end we must proceed. Enjoy!

TABLE OF CONTENTS

Foreword	iii
Preface	v
Table of Contents	vii
Chapter 1: Hydrogen Energy	1
Chapter 2: A Simple Hydrogen Engine Design	9
Chapter 3: Backfire In Hydrogen Engines Control Methods	17
Chapter 4: Water Induction In Hydrogen Engines	29
Chapter 5: Advanced Engine Conversion Direct Cylinder Injection	43
Chapter 6: Hydrogen Storage Compressed Gas And Liquid	51
Chapter 7: Hydrogen Stored As Metal Hydride The Safest Fuel On Earth	65
Chapter 8: Hydrogen Vehicle Prototypes	77
Chapter 9: Hydrogen Production And Electrolysis	103
Chapter 10: The Hydrogen Homestead	115
Chapter 11: The Hydrogen Fuel Cell Efficiency Is The Key	125
Chapter 12: Our Hydrogen Energy Future How We Begin	135
Appendix	
Hydrogen Equivalents	142
Unit Conversions	144
Bibliography	147
Index	165

Chapter 1

Hydrogen Energy

CHAPTER 1

HYDROGEN ENERGY

To me hydrogen energy is an emotional subject. I was just 14 years old when I watched a science teacher split water into its component parts, hydrogen and oxygen. The teacher collected the hydrogen in a small toy balloon. Then he dipped the string, which was tied to the balloon, in alcohol and set it on fire. The lighter-than-air balloon, filled with hydrogen, floated to the ceiling of the science classroom as the fire advanced steadily up the string. Then the explosion—a distinct detonation, a momentary flash of fire. The thing which stays with me most now, so many years after that experience, is what the teacher, Mr. Max Mitchell, did next. He went to the chalkboard and wrote:

Hydrogen plus oxygen yields water and ENERGY

The idea of fire resulting from the creation of water seemed like magic. Here was a better way to power the world. The by-product of burning hydrogen was pure water, which could then be electrolyzed again to make more hydrogen. The water could be used over and over and never be consumed. Best of all, there would be no air pollution!

I had loved science from my early youth and had made a determined decision in the third grade that someday I would be a scientist. Now I had a focus for my energy and enthusiasm—to bring to pass a hydrogen world. I became so excited about the idea that I went home and wrote a three-page report on the subject, which I submitted the next day to my science teacher. I really got carried away! Besides eliminating all of the pollution in the world, my report suggested using electricity from the car's alternator to generate the hydrogen while driving. (It was not until years later that I was introduced to the first law of thermodynamics and thereby discovered the flaw in my logic.) It took some serious effort to convince myself that such a "perpetual motion" car could not be a reality. However, the dream of the perfect, pollution-free car running on water-generated hydrogen never dimmed and is now, so many years later, finally becoming a reality.

In later years I came to have a broader understanding of what some authors call "the Hydrogen Energy Economy". Although hydrogen is the most abundant element in the universe, it is not normally found in its free or unreacted state in nature. It is almost always necessary to synthetically produce the hydrogen which we plan to use for the "Hydrogen Economy". (Naturally-occurring hydrogen has been discovered in vast underground deposits in eastern Kansas—just 150 miles from our Independence, Missouri, laboratory, but such an occurrence is rare and still has scientists baffled concerning its origin.) A discussion of how hydrogen is produced will be the subject of later chapters. For now, to establish a simple starting point, we can say that hydrogen is made by splitting water into its component parts—hydrogen and oxygen—which process utilizes energy.

Hydrogen, then, is not usually considered a source of energy, but rather is a form of stored energy. The energy to produce hydrogen from water can come from a wide variety of sources including solar, wind, hydroelectric, nuclear, natural gas reformation, and coal

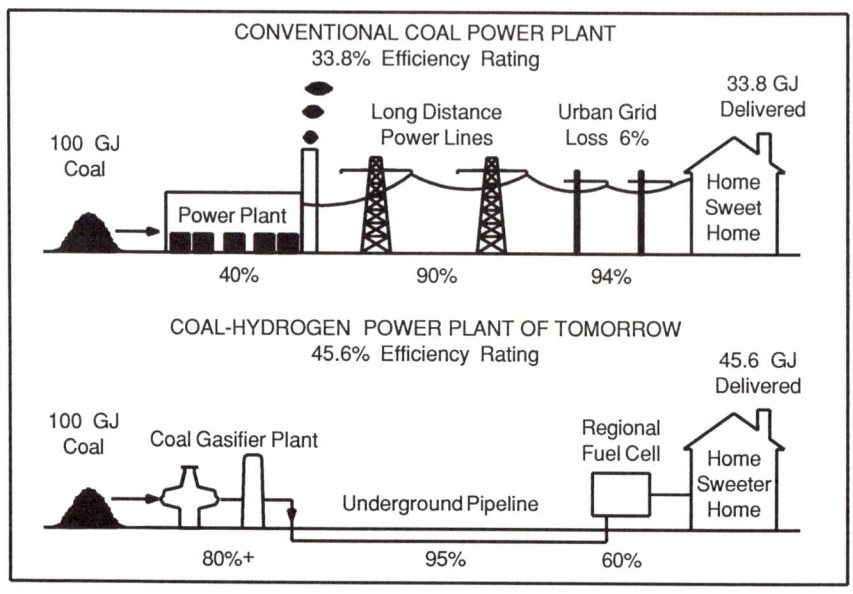

FIGURE 1-1: HYDROGEN FROM COAL ALTERNATIVE

gasification. When hydrogen has been synthesized from water using energy from one of these sources, it can then be stored, transported, and utilized in conventional energy consumption devices such as automobiles and gas appliances.

When hydrogen is utilized it is burned, which means it is combined with oxygen, usually from air, with the resulting by-product being the formation of water and the release of nearly as much energy as was required to synthesize the hydrogen in the first place. The reduction in the resulting energy liberated is what scientists refer to as the efficiency of the conversion cycle. In many cases, hydrogen's chemical properties make possible much higher utilization efficiencies than are possible with conventional fuels. In these cases, it is often possible to accomplish more useful work by converting the energy to hydrogen than by directly utilizing the primary fuel source. This phenomenon is illustrated in Figure 1-1 where the generation of electricity from coal and its distribution to the end-user is considered. The conventional method of accomplishing this task is to burn the coal to generate steam, which is then utilized to drive a turbine which produces the electricity. After considering the energy conversion and distribution efficiencies of this method, the net amount of energy delivered to the consumer is just 33.8 percent of the amount of energy actually produced by burning the coal. In addition to their low energy conversion efficiencies, coal combustion plants are also highly polluting and one of the principal sources of acid rain.

The better approach, which has yet to be utilized, would convert the coal into hydrogen through a process known as "coal gasification". The hydrogen could then be readily transported to urban centers of population via underground pipelines without high energy losses and without unsightly overhead high tension power lines which visually pollute some of the most beautiful spots on earth. Once in the city, the hydrogen is transformed into electricity by the use of highly efficient and non-polluting fuel cells—the net effect being that 45.6 percent of the energy from the coal would be delivered to the electrical consumer. This represents a 35 percent increase in power delivered to the consumer which translates into better resource utilization and a lower price to the user. Most significantly, the coal gasification process, which is already in commercial utilization as a source of hydrogen around the world, produces no sulfur dioxide, even when abundant high-sulfur coal is used. In hydrogen-from-coal plants, the sulfur is transformed into hydrogen sulfide ("rotten egg" gas) which is thermally decomposed at the plant into elemental sulfur and additional hydrogen. Pure sulfur is in high demand as a feedstock and is a naturally occurring, non-detrimental material. Everybody wins!

FIGURE 1-2: HYDROGEN PROPERTIES

Property	Value
Melting Point	-259.2°C
Boiling Point at 1 atm	-252.8°C
Density of Solid at -259.2°C	0.0866 g/cm^3
Density of Liquid at -252.8°C	0.0708 g/cm^3
Critical Temperature	-240.0°C
Critical Pressure	13.0 atm
Critical Density	0.0301 g/cm^3
Specific Heat at Constant Pressure	
Gas at 25°C	3.42 cal/(g)(°C)
Liquid at -256°C	1.93 cal/(g)(°C)
Solid at -259.8°C	0.63 cal/(g)(°C)
Heat of Fusion at -259.2°C	14.0 cal/g
Heat of Vaporization at -252.8°C	107 cal/g
Thermal Conductivity at 25°C	0.000444 cal/(cm)(cm^2)(sec)(°C)
Viscosity at 25°C	0.00892 centipoise

Encyclopedia of Energy, Lapedes, 1976.

A more impressive example of the advantage of converting conventional fuels into hydrogen is the reformation of natural gas into hydrogen to power fuel cell automobiles. More will be said about this later.

Another advantage of converting energy into hydrogen before utilization involves storability. In the case of solar energy, hydrogen can be produced during the day while the sun is shining and stored for use at night when there is no sun. At last it is really feasible to build a home which is completely powered by solar, because with hydrogen it is possible to store vast quantities of energy for times of need. For this reason, solar homes of the future will be able to "cut" themselves free from the "grid". In the case of solar automobiles, it is not possible to collect enough solar energy to power an automobile using just the area of the car's surface upon which to mount the solar collectors. It is possible, however, to collect the sun's energy on large, stationary collectors for conversion into hydrogen for storage. The hydrogen can then be used to refuel a hydrogen-burning automobile in a matter of minutes. Such a vehicle is capable of the range and operational performance we have come to expect.

In recent years, scientists have become concerned about a global warming trend referred to as the "greenhouse effect" which has now been traced to the release of carbon dioxide and other gases, such as methane, into the environment. Carbon dioxide is the principal by-product of the combustion of hydrocarbon fuels. Most of our energy utilization at present is based upon burning hydrocarbon fuels. This global warming trend is blamed for much of the severe weather which we are now experiencing. Assuming this is the case, we can only expect these problems to continue to grow in magnitude, causing destructive storms and serious droughts.

Looking at the situation from a more philosophical point of view, it is as though the earth were ill because of microscopic organisms pumping poisons into her systems. The earth has a SERIOUS viral infection destroying her vitality. MAN is that virus. The whole planet is going into convulsions because of man-made filth and pollution which are being continually dumped into the environment. By using clean, renewable energy sources and conversion to hydrogen, we can begin to reverse this deadly and dangerous trend. The problems which we face in this regard are serious and cannot be overstated.

In addition to using renewable energy sources to produce hydrogen, we must also continue our efforts to conserve energy consumption. Our initial steps of better insulated buildings and more fuel-efficient vehicles are not nearly enough to avoid widespread global calamity. It will also be necessary to develop new patterns of living which greatly reduce our dependence on hydrocarbon fuel sources and our resource wasting. Included in these societal changes, such things as using public transportation, living closer to work, and increasing domestic food production and storage will all be necessary. It doesn't even take a clever mind to extrapolate that our present course will eventually lead to unprecedented world suffering and sorrow.

Hydrogen, then, is not the "one" solution to all of our serious problems. However, hydrogen can, if properly supported, eliminate most urban air pollution, stop oil embargoes, improve trade imbalances, and, when implemented worldwide, reverse the trend toward global warming. Hydrogen is not the "one" solution to all of our complicated problems on this planet, but it is one giant step in the right direction. John K. Hansen, founder of Winnebago Industries, in speaking of hydrogen said, "I want to be highly involved in anything that is this important to the future of this great country." When the technology is right, and when millions begin to demand hydrogen energy, then we will begin to see this dream become a reality. In the next several pages I plan to report that the technology is now "right". The time to begin "hydrogen" has arrived.

Chapter 2

A Simple Hydrogen Engine Design

CHAPTER 2

A SIMPLE HYDROGEN ENGINE DESIGN

During the summer of 1963, at 15 years of age, I began my hydrogen career in earnest. I obtained a used gasoline-powered lawn mower, removed the engine, and began my first hydrogen engine experiments. I was in great hopes of demonstrating a hydrogen-fueled engine as a science fair project. I removed the liquid gasoline carburetor and replaced it with a gaseous fuel device of my own design, which "hydrogen carburetor" consisted of a large glass flask approximately one foot in diameter. In the top of the flask I placed a rubber stopper fitted with three different pieces of glass tubing—

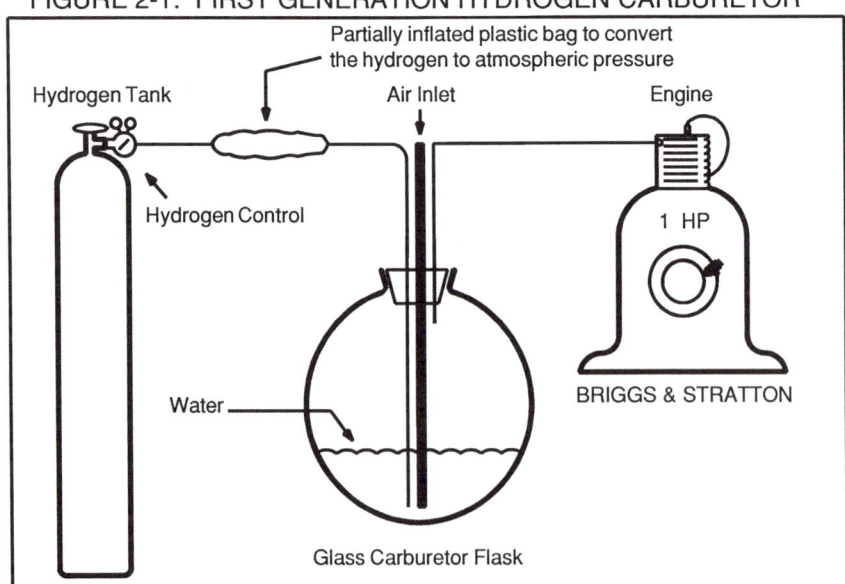

FIGURE 2-1: FIRST GENERATION HYDROGEN CARBURETOR

two long pieces and one short piece. (See *Figure* 2-1.) The flask was filled one-third full of water so that the two longer tubes reached down below the water line while the shorter tube only penetrated the top of the container. The two longer tubes were of different diameters. The larger tube, the air intake, had a calculated cross-sectional area equal to two-and-one-half times that of the smaller tube, the hydrogen intake. I selected this ratio because I knew I would need two-and-one-half volumes of air for each volume of hydrogen in order to get proper combustion. The short tube was hooked up to the engine. The idea was for the vacuum created by the intake of the engine to pull hydrogen and air into the flask in the correct ratio for combustion. Air would enter the flask through the large diameter tube while hydrogen would enter through the small diameter tube. The carefully selected diameters of the tubes would control the mixture of air-to-hydrogen as long as the hydrogen was at exactly atmospheric pressure. To convert the hydrogen to atmospheric pressure, I ran it through a plastic bag en route to the carburetor flask. The hydrogen, which was stored in a high pressure welding cylinder, had a small needle valve on the outlet. The idea was for my assistant (younger brother Lewis Billings) to open and close the hydrogen valve as needed to keep the plastic bag exactly one-half inflated with hydrogen. Since the outside air was pushing in on the bag, I determined that a half-full bag would mean that the two gases would be at exactly the same pressure. Based on the outcome of the experiment I suppose that the mixture must have been very close to the chemically correct ratio (stoichiometry).

It took some convincing to get my mother to sign for the hydrogen cylinder I needed, but, after all, it was "in the name of Science". The location selected for the experiment was our back porch patio. When all of the apparatus had been carefully prepared, research assistant Lewis had been trained on the operation of the hydrogen valve, and the rope had been coiled around the lawn mower engine for starting, I paused to do a "last minute" system check. Having had a serious Christian upbringing, I decided that a short but sincere word of prayer would be an appropriate first step before performing the experiment. As we prayed, I became concerned about the large glass flask. After the prayer, I went to the garage and found my father's sturdy flying jacket, which we zipped around the flask, tying the top and bottom with a rope to make sure it would hold. Then, after a NASA-style countdown from ten, I pulled the rope for all I was worth. The engine began turning in response to the pull, but gradually slowed down without firing, as though it was getting no fuel. Then as it came to a stop, the crankshaft turned backwards the famous one-half turn. During the backwards half turn, a spark inside the engine ignited the intake chamber and the flame progressed rapidly back through the tubing into the glass carburetor flask. At that moment, there was a terrible explosion, which brought my mother running from inside the house. The jacket was destroyed. The only part of the flask remaining intact was the neck, and pieces of glass were scattered over a wide area. However, neither I nor my brother Lewis suffered a scratch in spite of the fact that he was only inches from the flask. (My father has not piloted a private plane since that day.)

It took more than a year to convince my mother that another hydrogen experiment was in order. In 1964, the Utah Academy of Science, Arts, and Letters sponsored a contest for high school students in which the prize was "four hours of consulting on your science project from the college professor of your choice". I won the prize and chose a Mechanical Engineer from Brigham Young University as my consultant. When I explained to my consultant what I had in mind, he was somewhat skeptical about my choice of projects. (This was despite the fact that I failed to mention to him the disastrous results of my experiment one year earlier.) He told me that it was impossible to run an engine on hydrogen. Pulling a well-worn textbook from his shelf entitled, Gas Engine Designs by C.E. Lucke, published in 1905, he read the following: "Gases rich in hydrogen pre-ignite easily and approximately one atmosphere should be deducted from the compression pressure allowable with no hydrogen, for each five percent of hydrogen present." (In another article published in the Report of the Empire Motor Fuels Committee, 1924, H.R. Ricardo said, "If, however, an attempt was made to run with a rich hydrogen-air mixture, violent pre-ignition occurred, accompanied by firing back into the carburetor which rendered further running impossible".) My advisor then suggested several other good ideas for science projects that might be more successful. That experience ended my hydrogen energy career for the rest of my junior year of high school.

By the time I entered my senior year of high school, my interest in hydrogen energy was renewed. This time I decided to couple the gasoline lawn mower engine to an electric motor which would allow me to motor the engine while I was attempting to refine my carburetion system. My carburetion system evolved into the simple but effective design shown in *Figure* 2-2. A small hydrogen line introduced pure fuel into the intake chamber immediately above the intake valve. While the intake valve was closed during the non-intake parts of the cycle, the hydrogen would flow constantly out of the small tube, displacing the air in the large tube and making a pocket of nearly pure hydrogen in the area surrounding the intake valve. When the intake valve opened, the pocket of hydrogen was drawn into the combustion chamber first, with the remaining space then being filled with air. Inside the combustion chamber the hydrogen and air were not very well mixed, creating a charge stratification phenomenon which greatly slowed down and im-

14   The Hydrogen World View

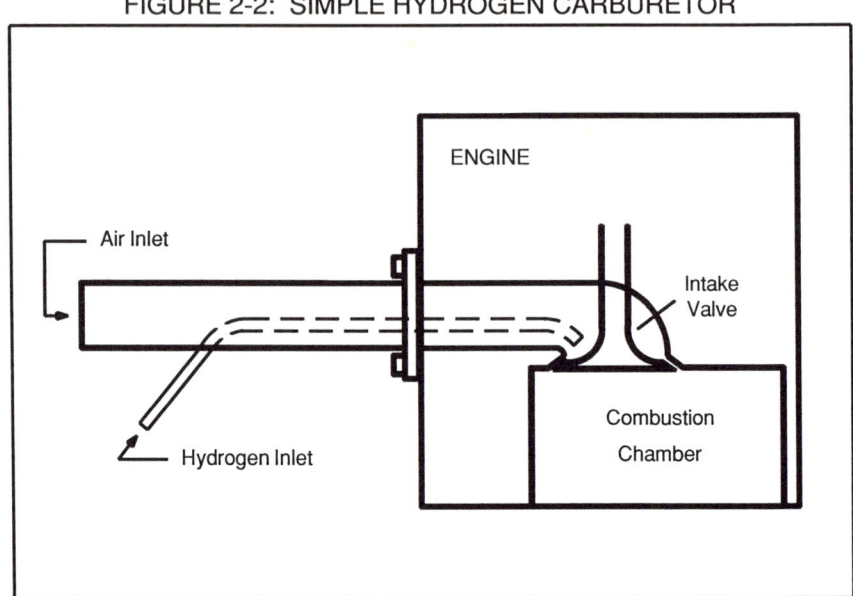

FIGURE 2-2: SIMPLE HYDROGEN CARBURETOR

proved the hydrogen combustion. This carburetion system worked flawlessly from the very first.

At this point it was time for the big test. Having the tremendous success of the lawn mower engine behind me, I confronted my father at the breakfast table one morning. I declared that science was definitely marching forward. I told him about the lawn mower experiment and brought up the brand new Chevrolet we had just bought. Then I popped the big question: "How would you like the new Chevy to be the world's first hydrogen car?!"

He said he needed some time to think about it. However, when you are at a point of scientific breakthrough, you just can't wait around! So, during my first class in school that morning I talked to my friend Lynn Barker about using his Volkswagen for an experiment. He said no.

I tried to convince him. I told him it was for science . . . it was in the name of science! I told him he would be the famous owner of the world's first hydrogen car! Then I took him up to the lab and showed him the lawn mower engine. Finally, he said O.K.

That night we tore the carburetor off his engine. I made one of my special "hydrogen" carburetors in the metal shop that would fit a Volkswagen, and I was just starting to hook it up when my father walked around the corner of the science building. He asked what we were doing. I told him "Research."

"No you're not," he said. "If you're going to blow up anybody's car, it's going to be ours." (Now my father really had no basis for jumping to that kind of a conclusion about my work--at least he had hardly any basis.) I was excited. I thought he was talking about the new Chevrolet. When I found out he was talking about the old Model A, I was only a little bit disappointed, but, unbeknown to me at the time, using the Model A was a significant choice.

My first task in converting the Model A was the carburetor. I designed one similar to the one for the lawn mower except, of course, I made it larger. As in my previous experiments, the system performed satisfactorily except for an occasional backflash.

In the Model A prototype, two compressed gas cylinders containing hydrogen were installed in the back of the vehicle and manifolded together. In the initial design, I used an "LBL"-type throttle control valve (Little Brother Lewis). I would signal thumbs up, and he would open the valve for more power. Thumbs down meant less power. It may not sound too sophisticated, but operating in this manner we drove all over town.

To prepare for the science fair, I collected samples of the exhaust from the hydrogen engine and had them run through the university's gas chromatograph to verify the tremendous advantages of a pollution-free fuel. To my surprise and dismay, the results indicated that a substantial amount of nitric oxide was present in the exhaust of the hydrogen engine. After a little research I learned that nitric oxide is formed when nitrogen and oxygen (the two constituents of air) are heated, as in an internal combustion engine. Since hydrogen has a higher flame speed than gasoline, the peak temperature in my engine was higher than in a gasoline-fueled car. As a result, the exhaust from my Model A engine converted to hydrogen had a much higher concentration of nitric oxide than did the exhaust from the engine operating on gasoline. (Of course my car had no carbon monoxide, unburned hydrocarbons, or carbon dioxide as was the case when the vehicle operated on gasoline.) But if hydrogen was to become the pollution-free fuel of the future, some method of eliminating the nitric oxide formation would have to be discovered.

Although I did not have significant problems with backfire in this prototype, I learned subsequently that it is usually a very serious problem when converting engines to operate on hydrogen. By virtue of some unseen guiding hand or what some might call "luck", the first two engines which I converted were both "L-head"-type engines. L-head engines have a very high surface-to-volume ratio which drastically reduces problems with hydrogen backfire. Realizing now that no current-production automobiles back then utilized L-head engine designs, I see the significance in my having "chosen" two L-head engines—the Model A Ford and the Briggs and Stratton—as my first two prototypes.

Chapter 3

Backfire In Hydrogen Engines Control Methods

CHAPTER 3

BACKFIRE IN HYDROGEN ENGINES CONTROL METHODS

I often meet people who claim to have operated a car on hydrogen. My standard response is a question: "What backfire control technique did you use?" If they can properly answer that question, I know I am talking to a genuine colleague. If they say they had no problem with backfire in their engine, I become immediately skeptical about their claim.

Backfire or backflash, as some people prefer to call it, is a serious problem associated with hydrogen engines that must be taken into consideration when designing a conversion system. In the early days of hydrogen engine research, it was very difficult to isolate the cause of this problem because the data was often contradictory and not repeatable. Now, several separate mechanisms have been identified as independent mechanisms causing the backfire problem.

To begin to understand the backfire problem, we should first understand some of the chemical properties of hydrogen. Hydrogen has a very fast flame speed. To put it in engineering "jargon", the laminar flame speed of hydrogen in air is an order of magnitude faster than that of hydrocarbon fuels. What this means is that if two pipes 100 feet long were placed side by side—one filled with hydrogen and air, and the other with a hydrocarbon fuel and air—

and then ignited at exactly the same time, the hydrogen flame would burn through the pipe, arriving at the far end in one tenth the time it would take the hydrocarbon-fuel flame to make the same trip. Although there is a difference in flame velocity from hydrocarbon fuel to hydrocarbon fuel, compared to hydrogen they are all relatively equivalent.

Another very important characteristic of hydrogen is its ignition parameters. The energy required to ignite hydrogen is an order of magnitude less than that required to ignite hydrocarbon fuels, but that energy must be supplied at a higher temperature. This statement may seem confusing, so I will explain. If a fuel/air mixture were placed in a carefully controlled oven and the temperature gradually increased, a temperature would eventually be reached where the fuel would ignite. This temperature is referred to as the "ignition temperature" and varies from fuel to fuel. In such an experiment the oven would be heated to a higher temperature to achieve the ignition of hydrogen than it would for the ignition of hydrocarbon fuels. The ignition temperature for hydrogen is 958°C (1756°F) compared to 820°C (1508°F) for gasoline and 795°C (1463°F) for methane.

In another experiment, a mixture of air and the fuel to be tested is placed in a reaction chamber which has been equipped with a spark plug. For this experiment, the spark plug system is specially designed so that very weak sparks can be discharged across the spark plug gap. From such experiments we learn that hydrogen ignites with a very weak spark while gasoline requires a spark of ten times more energy before igniting.

To understand this better, it is important to realize that a spark of the smallest magnitude creates a temperature greater than the ignition

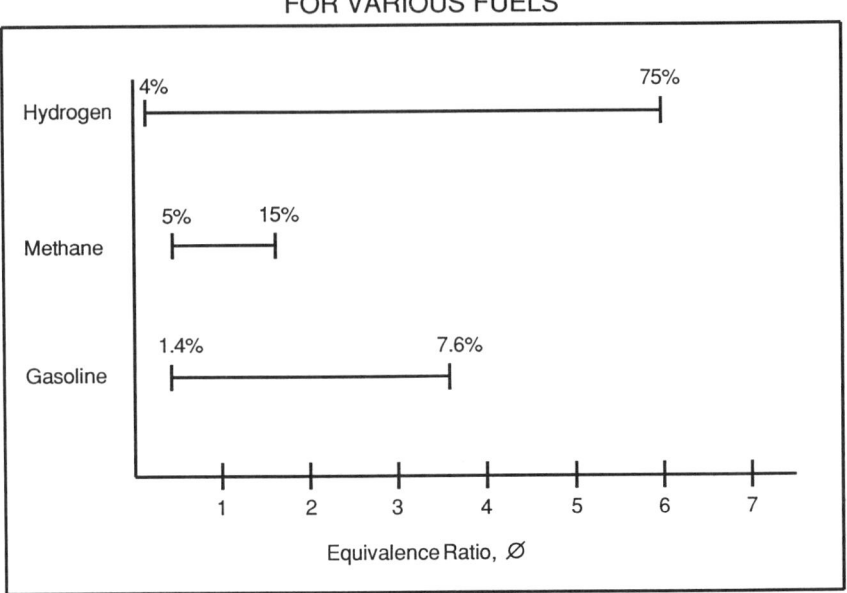

FIGURE 3-1: AIR-FUEL FLAMMABILITY RANGES FOR VARIOUS FUELS

temperature for hydrogen or hydrocarbons, but the amount of high-temperature energy supplied by a very weak spark is not sufficient to initiate hydrocarbon fuel combustion. The ignition energy for hydrogen is 457 joules, whereas the ignition energy for methane is 4,617 joules. Hydrogen requires a higher temperature for ignition, but only one-tenth as much energy need be supplied at that higher temperature as compared to the hydrocarbon fuels.

Another unusual chemical property of hydrogen as compared to hydrocarbon fuels is its flammability range. On the lean end, a mixture consisting of 4 percent hydrogen and 96 percent air is flammable. On the rich end, a mixture consisting of 75 percent hydrogen and 25 percent air is still flammable as is any mixture within that range. By contrast, methane is flammable in concentrations of only 5 to 15 percent in air. What this means is that if a combustion chamber was filled with 25 percent hydrogen and 75 percent air, it would ignite and be combustible. If the same chamber was filled with 25 percent methane and 75 percent air, it would not burn. A comparison of the flammability ranges for various fuels is presented in *Figure* 3-1. The broad flammability range of hydrogen is a very important consideration in the design of hydrogen energy systems.

Having discussed these basic chemical properties of hydrogen, we can now begin to understand the backfire problem. To the driver of a hydrogen automobile, "backfire" is the common name for the loud explosion that occurs randomly in a poorly-designed engine conversion. (Actually, the occurrence of backfire is not completely random. It usually occurs during a press conference, when the vehicle is being filmed by a TV crew, or when you are letting a very important visitor test drive the car.) To a researcher, backfire is the occurrence of one or the other (or sometimes both) of two distinguishable phenomena. A true backfire occurs when an explosive mixture of fuel and air builds up in the exhaust system of the engine and is then ignited, with the explosion coming out the tailpipe of the vehicle. This type of backfire is common in poorly-tuned engines running on gasoline and is fairly rare in hydrogen engines. The second type of backfire, which is more properly referred to as "backflash", is that phenomenon which occurs when a combustible mixture of hydrogen and air in the intake system of the engine is ignited. In this case the explosion comes out of the carburetor of the vehicle, very often damaging the equipment. Backflash is a very common problem in hydrogen engines. Let's begin to list the reasons why.

22 The Hydrogen World View

FIGURE 3-2: INDUCED IGNITION IN HYDROGEN ENGINES

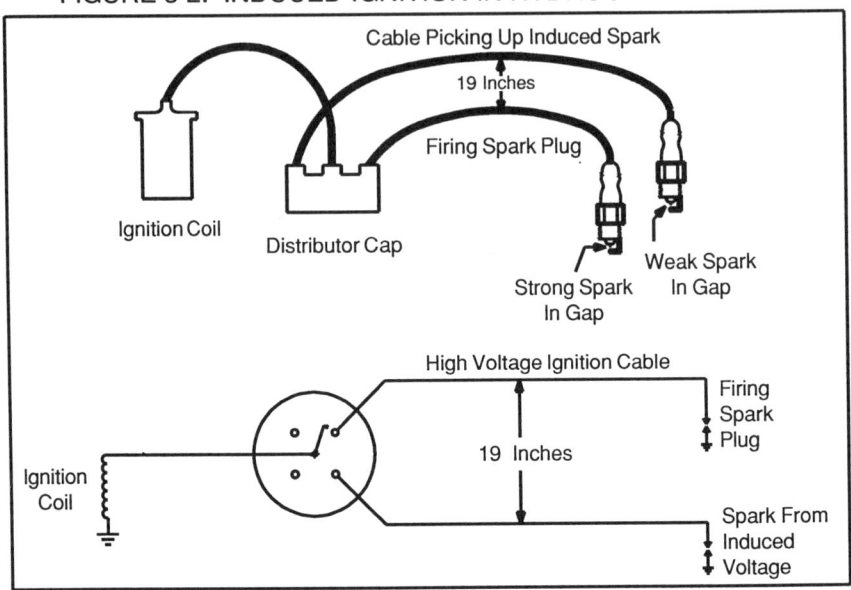

INDUCED SPARK IGNITION

In a current-model, high-performance engine, fuel is ignited at the appropriate time in 4, 6, or 8 alternative cylinders. The ignition of the fuel is usually accomplished by means of a high voltage ignition system which is directed to the cylinder ready to be fired by a distributor cap and a rotor. Careful experimentation on a conventional ignition system reveals some interesting phenomena. When the ignition system fires a spark in the proper cylinder, simultaneously a much smaller "induced" spark is created in other cylinders. These induced sparks are usually caused by the electromagnetic interference between the cables running from the distributor cap to each individual spark plug. In laboratory bench tests, we found that an induced spark could be reliably reproduced at the end of an ignition cable located as far as 19 inches away from the cable where the primary spark had been fired. (See *Figure* 3-2.) These induced sparks are sometimes present in a cylinder during the intake stroke and come in direct contact with a flammable mixture of fuel and air. In a conventional gasoline engine, these induced sparks are too weak to ignite the gasoline/air mixture. In the hydrogen engine, however, it is a different story. Since, in this case, only one-tenth as much energy is required for ignition, the hydrogen/air mixture ignites readily. If the cylinder is in the intake cycle, the flame burns up through the intake valve and into the intake manifold causing a backflash, much to the chagrin of the vehicle operator and the amusement of the press.

The induced spark problem can be minimized and effectively eliminated by employing one or a combination of the following techniques:

A. The spark gap in the spark plugs can be reduced. The peak voltage of an ignition system will rise as high as necessary to achieve gas ionization. Once ionization occurs, the spark gap becomes almost a direct short and the voltage falls off. By reducing the spark gap, the amount of air needing to be ionized is reduced, thereby lowering the peak voltage of the ignition system. The strength of an induced spark is directly proportional to the voltage of the primary source. For this reason, a reduction in the peak voltage of the ignition system by reducing the spark gap can greatly reduce the tendency for backflash.

B. Another simple but very effective method of preventing induced sparks is to run spark plug wires in a perpendicular fashion rather than in the traditional, neat, parallel fashion. When wires cross each other at a 90-degree angle, there is no resulting induced spark. Wires can be held in place with cable ties. The proper implementation of this method is depicted in *Figure* 3-3.

C. A third but perhaps more exotic way of reducing induced spark is by using shielded ignition cables. Shielded ignition cables by themselves, however, will not solve this problem. The shields must be properly grounded to the engine in order to be effective. Shielded ignition systems such as those employed in aircraft are one good way to implement this technique.

FIGURE 3-3: INDUCED IGNITION CAN BE MINIMIZED BY ROUTING CABLES PERPENDICULARLY

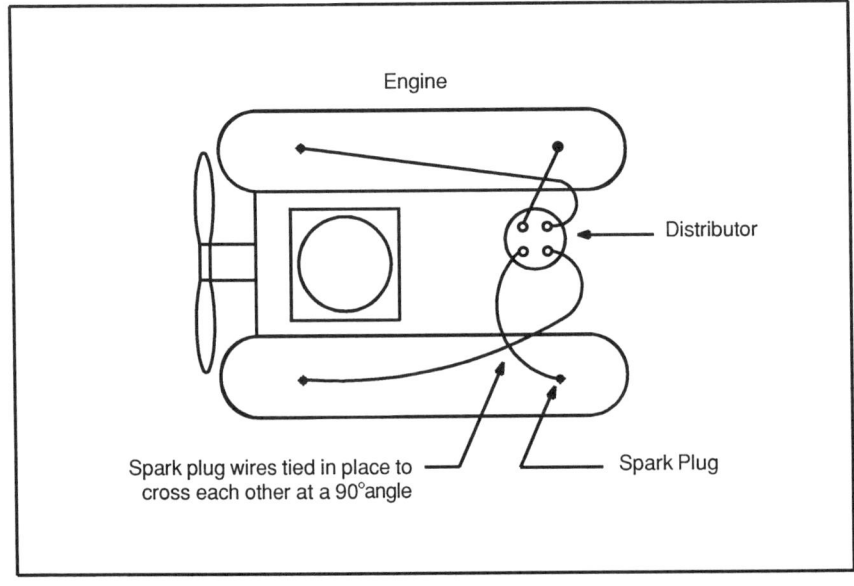

D. Induced spark can also be eliminated by dedicating an independent ignition source to each spark plug and making the run between each coil and its individual spark plug very short.

E. The final method which I will suggest for preventing induced sparks is the utilization of single cylinder engines. Obviously, this method of control has limited application, but, when applicable, it is the most effective method of all. The induced spark problem does not exist in single-cylinder engines.

Having tried all of the above techniques for induced spark control, I have achieved my best results with a combination of the first two recommendations. I reduce the spark gap to reduce the overall voltage and then route the wires to run at 90-degree angles. An interesting experiment to test the effectiveness of this backflash control method is to attempt to operate a multi-cylinder engine with the cables routed neatly in parallel paths and then to switch them to the 90-degree angles. When the engine is operated in a load condition, repeatable results are obtained.

PRE-IGNITION SOURCES

Another significant cause of backflash in the hydrogen engine is from pre-ignition sources. An interesting demonstration of this phenomenon is shown by placing a hydrogen engine in a test laboratory and connecting it to a dynamometer so that it can be tested under normal operating loads as though it were powering an

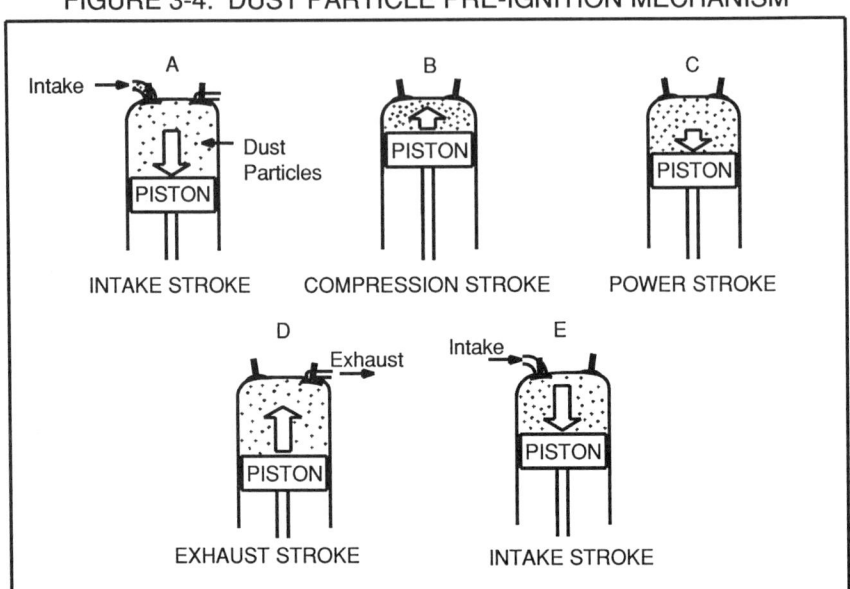

FIGURE 3-4: DUST PARTICLE PRE-IGNITION MECHANISM

automobile. The intake filter is removed and a small quantity of very fine dust particles allowed to go into the engine intake. This experiment will create no extraordinary results if the engine is operating on gasoline, but when the engine is operating on hydrogen, the result is an almost immediate backflash. The reason is illustrated in *Figure* 3-4.

The particles are drawn into the combustion chamber during the intake stroke (A). The fuel is then compressed (B), and the charge is ignited, resulting in the power stroke (C). During the exhaust stroke the exhaust valve opens and the burnt charge, along with most of the particles, is pushed out into the exhaust system (D). The cycle repeats itself with the intake valve opening and a new intake stroke beginning (E). At this point, however, some of the dust particles, which are still hot from the last cycle, remain in the combustion chamber. As the hydrogen fuel comes into contact with these hot particles—which have enough thermal mass to ignite the hydrogen—the incoming fuel charge is ignited and burns quickly back into the intake manifold, resulting in the detonation. In the case of the hydrocarbon engine, these small particles do not have sufficient thermal mass to ignite the hydrocarbon fuel, so this is not a problem for the gasoline-powered car.

Other sources of pre-ignition backflash are carbon buildup within the combustion chamber (caused by operating the engine on hydrocarbon fuels) and rusting spark plugs. In the case of carbon buildup, the carbon becomes hot during engine operation. This hot source, if at a sufficiently high temperature, can ignite the incoming charge of hydrogen and air during the intake stroke causing backflash. The case of the rusting spark plug is even more interesting. Rust, or iron oxide, is an excellent catalyst for the ignition of hydrogen at high temperatures. As a result, the incoming hydrogen/air charge comes into contact with the hot spark plug which provides a source of catalytic ignition and the resulting backflash.

To prevent these problems, several things can be done. First of all, in the case of carbon buildup, operating an engine on hydrogen will quickly "burn out" any carbon deposits left over from gasoline operation. This leaves the hydrogen expert a choice: either tear the engine apart and clean it or tolerate the problem for several minutes of operation and hope it will go away. (It usually does.) In the case of the rusty spark plugs, the best solution is to swap out conventional spark plugs for plugs made of non-corrosive material such as stainless steel. Stainless steel spark plugs for most engines can be obtained through Champion Spark Plug Company at a fairly nominal cost.

These examples of backflash-causing phenomena do not constitute the entire list of possibilities, but are sufficient to give the researcher a handle on the parameters that must be controlled for a successful conversion. When a backflash does occur, it often causes damage to the hydrogen fuel system and is sufficiently loud to deter a would-be enthusiastic driver.

Fortunately, in addition to the above methods of eliminating backflash in the hydrogen engine, there are a couple of other global control techniques for eliminating backfire or backflash. These methods have been tested and found to reduce or eliminate backflash in a wide diversity of engines. At the same time, many of these techniques for controlling backflash are effective in substantially reducing or even eliminating, for all practical purposes, the generation of nitric oxide as a pollutant in the exhaust. These methods will be discussed in the next two chapters.

Chapter 4

Water Induction in Hydrogen Engines

CHAPTER 4

WATER INDUCTION IN HYDROGEN ENGINES

One interesting method of controlling backflash and a host of other problems associated with hydrogen engine conversions is water induction. This is the method of engine conversion which I utilized in 1972 to convert a Volkswagen Superbeetle to hydrogen. That vehicle won first place for emissions, representing Brigham Young University in the National Urban Vehicle Design Competition held at the General Motors Proving Grounds near Detroit, Michigan.

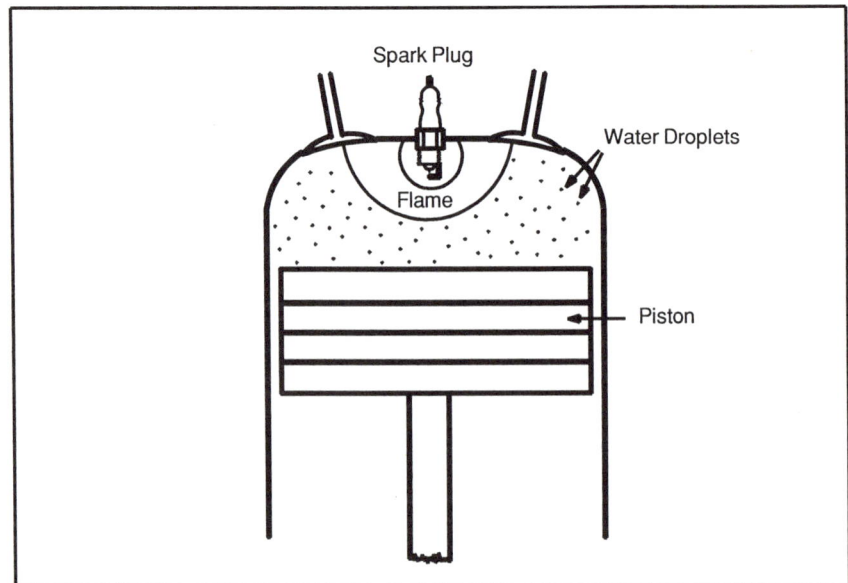

FIGURE 4-1: WATER INDUCTION MECHANISM

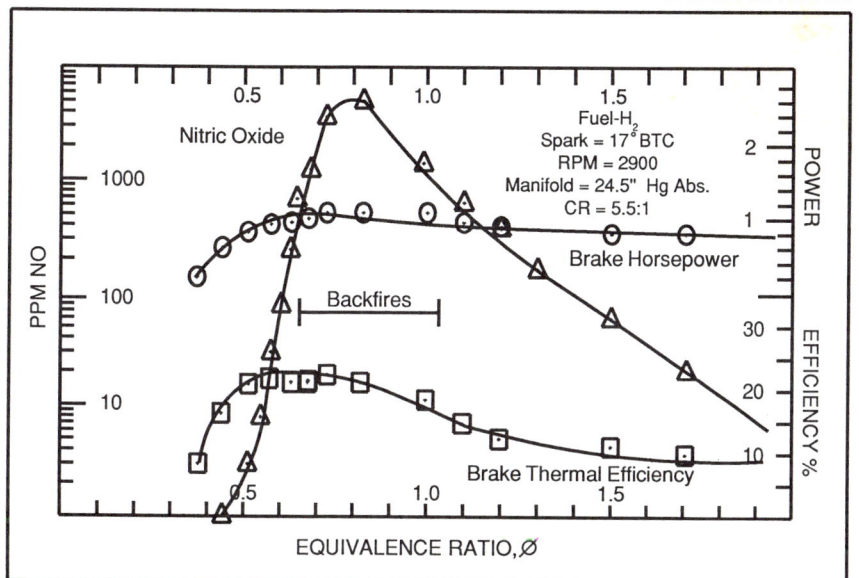

FIGURE 4-2: NITRIC OXIDE FORMATION IN A HYDROGEN ENGINE

The technology employed in water induction is very interesting. It is accomplished by inducting tiny droplets of water (not vapor) into the combustion chamber with the air/fuel charge. When the charge is compressed and ignited, the tiny droplets of water become involved in the ensuing chemical reaction, causing a multiplicity of results. Some of the results are very beneficial and even surprising.

To understand this technology, consider the combustion chamber depicted in *Figure* 4-1. Here, the charge inside the combustion chamber has been ignited by the spark plug and is beginning to propagate through the combustion chamber. Outside the arc of flame, the air/fuel mixture is filled by a mist of very tiny droplets of water. As the flame front encounters these tiny droplets, it instantly transforms them from liquid water into water vapor or steam. This phase change causes the water to expand 1400 times. In the process, a significant amount of heat energy is consumed. Now, consider these important ramifications: Since the peak combustion temperature is substantially reduced, auto-ignition sources are not heated to a high enough temperature to cause the auto-ignition phenomenon to take place. For this reason, water induction virtually eliminates all backflash from this source.

Water induction is also very effective in reducing the formation of nitric oxide pollution. As you will recall from an earlier chapter, nitric oxide is formed when nitrogen and oxygen, the two constituents of air, are heated to high temperatures. The formation of nitric oxide in an uncontrolled hydrogen engine is presented in *Figure* 4-2 as a function of hydrogen equivalence ratio. If a water induction system is installed on the engine and adjusted to provide sufficient droplets of water so as to maintain the peak combustion temperature below the nitric oxide formation threshold of 1300 degrees Centi-

FIGURE 4-3: EFFECT OF WATER INDUCTION ON THE FORMATION OF NITRIC OXIDE

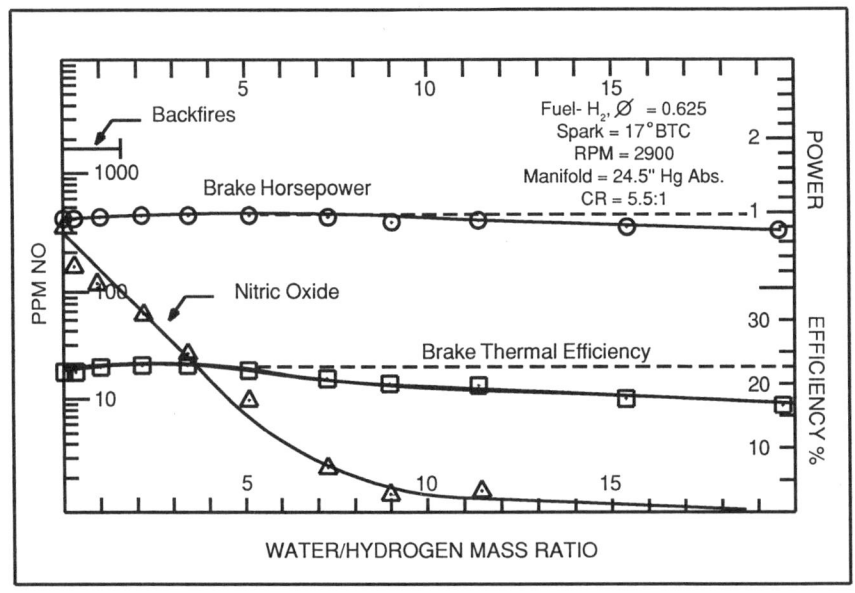

grade (2372°F), this undesirable chemical reaction, as shown in *Figure* 4-3, is virtually eliminated. As a result, the nitric oxide pollution coming from a hydrogen engine with water induction properly installed is one part per million or .02 grams per mile.

As I mentioned earlier, in 1972, my colleagues and I utilized the water induction technique in the vehicle that won first place for emissions in the Urban Vehicle Design Competition (UVDC). That meant that our little Volkswagen was less polluting than any other car in the competition.

At this point an interesting story is warranted. The intercollegiate competition to build a pollution-free car for the UVDC had resulted in numerous interesting entries. Many of the cars were powered by electric batteries. Since there was no exhaust, they were all given perfect scores—representing zero air pollution. Another entry, from a university whose name I will leave anonymous, utilized a high-speed flywheel as a mechanism for storing energy. Before each test, this car was pushed up to a large diesel tractor engine which was then coupled externally to the flywheel. As the flywheel stored up its energy, the diesel engine spewed out a noticeable cloud of black smoke. Then, utilizing energy stored in the flywheel, the car was driven inside the test laboratory where technicians certified it to be pollution free with a perfect score of zero.

To really appreciate this story, it is necessary to understand the scoring formula used by the judges. Three categories of pollutants were monitored: carbon monoxide, unburned hydrocarbons, and total oxides of nitrogen. Sophisticated General Motors Proving Grounds instruments were used to measure the concentration of each of these pollutants. The concentration of each pollutant was then multiplied by a weighing factor and the result for each added together to determine the overall score for a particular entry. The lower the score the better. Since the electric and the flywheel vehicles had a perfect score of zero, the most we could hope for with the hydrogen Volkswagen was a tie. Our hopes were somewhat dashed, however, when we learned that the test-monitoring instruments at the Proving Grounds were so sophisticated that they could actually measure the pollution in ambient air. Although our water induction method virtually eliminated nitric oxide formation, we were quite certain that they would be able to detect the very small amount of nitric oxide which, in fact, was generated inside our engine.

Shortly before our turn to be tested, the UCLA entry was tested. It was also a hydrogen-fueled vehicle which used exhaust gas recirculation as a method of controlling nitric oxide. Although the UCLA score was very impressive, their nitric oxide emission was sufficiently high to put them far behind the perfect scores of the electric and flywheel vehicles. Nevertheless, they took the lead for vehicles with internal combustion engines in front of entries fueled by methanol, ammonia, natural gas, and propane.

When the Brigham Young Superbeetle was tested, an interesting thing happened. Minute quantities of carbon monoxide and unburnt hydrocarbons in the surrounding air were drawn into the combustion chamber of the hydrogen engine and burned. As a result, our score for unburned hydrocarbons and carbon monoxide were both negative numbers. We did generate a very insignificant amount of nitric oxide which the engineers measured to be 0.022 grams per mile on the EPA-4 test cycle. When the final score was added up, our entry with the hydrogen engine and water induction ended up with a negative total! We removed more hydrocarbons and carbon monoxide from the air than the quantity of nitric oxide that we produced. One of the test technicians from General Motors made the pronouncement, "This car cleans the air as it drives." Immediately, nationwide attention focused on the hydrogen car from Brigham Young. News clippings and letters started pouring in from all over the United States. (The most interesting article came from the Los Angeles Times. The headline read, "This car would probably stall in Los Angeles air".)

The water induction technique had other benefits. For example, there was an increase in brake thermal efficiency of the engine as water induction was applied. We also observed a corresponding increase in horsepower output of the engine. These results of applying water induction to a hydrogen engine are given in *Figure 4-3*. At first I was somewhat reluctant to accept these test results. Usually when you cool down the peak combustion temperature of an engine, you pay the price of a decrease in horsepower output and engine efficiency. In this case, the reverse was true. This phenomenon can best be understood by considering events transpiring inside the combustion chamber on a microscopic level. As the propagating flame encounters the tiny water droplets, they are vaporized into steam and expand dramatically. (Refer to *Figure* 4-1). Although the vaporizing of the water droplets reduces the peak combustion temperature, the amount of energy consumed as the water undergoes a phase change generates a greater increase in pressure than would result if that same energy were utilized to heat and expand the air and combustion products normally encountered in the combustion chamber. In other words, we have some of the benefits of a Rankine-cycle device or a steam engine. After all, it is the pressure of the gases which force the piston down and not temperature. If a high pressure can be achieved inside the combustion chamber without high temperature, then all the better. Such is

36   The Hydrogen World View

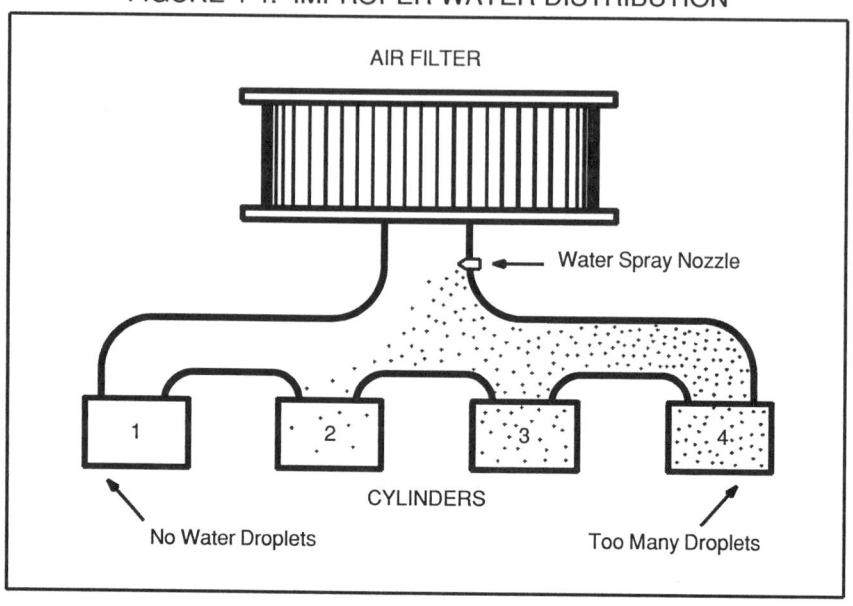

FIGURE 4-4: IMPROPER WATER DISTRIBUTION

the case with a well-designed water induction system on a hydrogen engine.

Like all real-world systems, however, there are some problems with water induction. One difficult problem is how to accomplish the water induction itself. In order to be effective, the tiny water droplets must be uniformly dispersed throughout the combustion chamber. Any pockets of combustible product within the chamber devoid of the tiny water droplets will generate high combustion temperatures, nitric oxide, and possibly even cause backflash. Getting the tiny water droplets uniformly distributed within the combustion chamber is a "ticklish" task.

Figure 4-4 shows a typical intake manifold system with a water atomizing spray nozzle on the right side of the common air inlet. As can be seen, the stream of incoming air sweeps the water droplets into the intake manifold, with the majority of them ending up in cylinder number four. A similar nozzle installed over the inlet to each cylinder has also been tested. In this case, the droplets were evenly distributed between each of the cylinders, but the nitric oxide formation results would indicate that there were pockets within the cylinders devoid of the water droplets because nitric oxide concentrations averaged 100 to 500 times higher than in better-designed systems.

One very good way to accomplish water induction is to use the "incredible" and well-developed technology of a good gasoline carburetor. I have had my best results with Holley carburetors. The problem of uniform liquid distribution is something automotive engineers have dealt with for years. In a gasoline engine, if the fuel is not properly distributed within the combustion chamber, the end result is incomplete combustion and very high levels of air pollution. Consequently, gasoline carburetors are designed to put a constant ratio of liquid-to-air into the chamber at all engine speeds, and to disperse the liquid as evenly as possible.

For my experiments I hooked up my inlet water line to the gasoline carburetor. In order to get enough water it was necessary to remove and to drill out the carburetor orifices to a size approximately one-third larger than were found in the stock carburetor. The actual orifice size for a given engine should be determined by measuring the amount of water consumption verses the amount of hydrogen consumption. This data can then be compared to *Figure* 4-3 to see if the amount of water induction is what it needs to be. If necessary, the orifices can be drilled out more to increase the water induction, or new orifices can be purchased to decrease the amount. For a hydrogen carburetor in this type of system, my preference is an Impco propane carburetor manufactured by the Impco Carburetor

FIGURE 4-5: GASOLINE CARBURETOR UTILIZED TO ACCOMPLISH WATER INDUCTION IN A HYDROGEN ENGINE

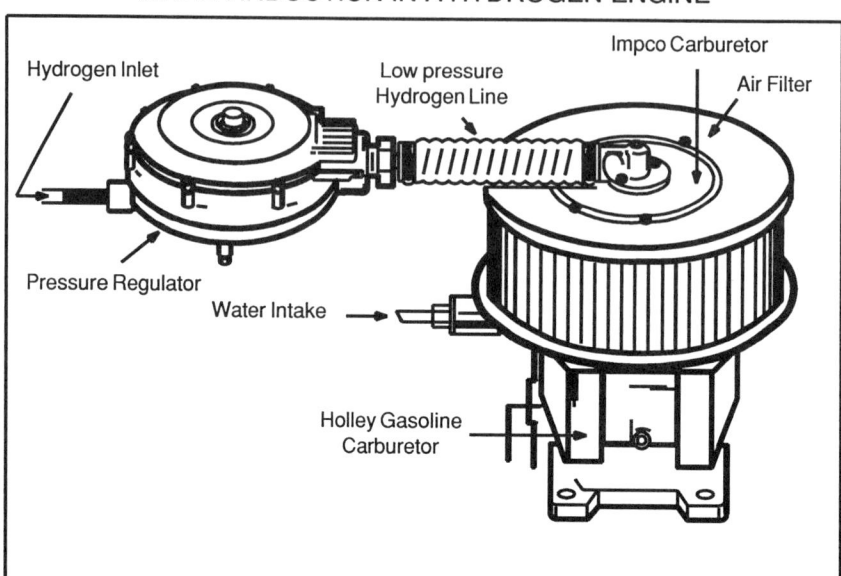

Corporation of Cerritos, California. They build a gas mixer which sits piggyback on top of the regular gasoline carburetor. In propane conversions, this piggyback configuration affords a dual-fuel capability for the automobile. In my case, I used the propane carburetor to mix the air and hydrogen, and the gasoline carburetor to induct the water. A schematic of this system is outlined in *Figure* 4-5. The regulator shown in the figure reduces the hydrogen/gas pressure to that pressure required by the Impco carburetor. This regulator is extremely important for proper operation of the system, and I would highly recommend that anyone testing such a system consider purchasing the regulator from Impco. Because of the very low pressures involved and the relatively high volume flow rates, the only other regulators which I have found to be suitable in this application are those manufactured for low pressure natural gas regulation.

To modify the Impco carburetor, it is necessary to compensate for the fact that hydrogen has a lower energy density per unit volume than does natural gas or propane. The first step in making this adjustment involves a modification to the pressure regulator. Impco supplies different springs to convert utilization of the regulator from propane to natural gas. For hydrogen, the system seems to work best if the regulator is operated at atmospheric pressure, which is achieved by completely eliminating the spring. This is done by removing the four small screws on the top of the regulator which releases the cover and allows access to the spring. After removing the spring, the cover should be reinstalled with the four screws. (Refer to *Figure* 4-6.)

Next, it is necessary to make the fuel mixture adjustments on the Impco carburetor itself. There are two adjustments—the idle mixture adjustment and the full power adjustment. It is best to set the idle mixture as lean as possible while still maintaining smooth

FIGURE 4-6: IMPCO PRESSURE REGULATOR MODIFICATIONS

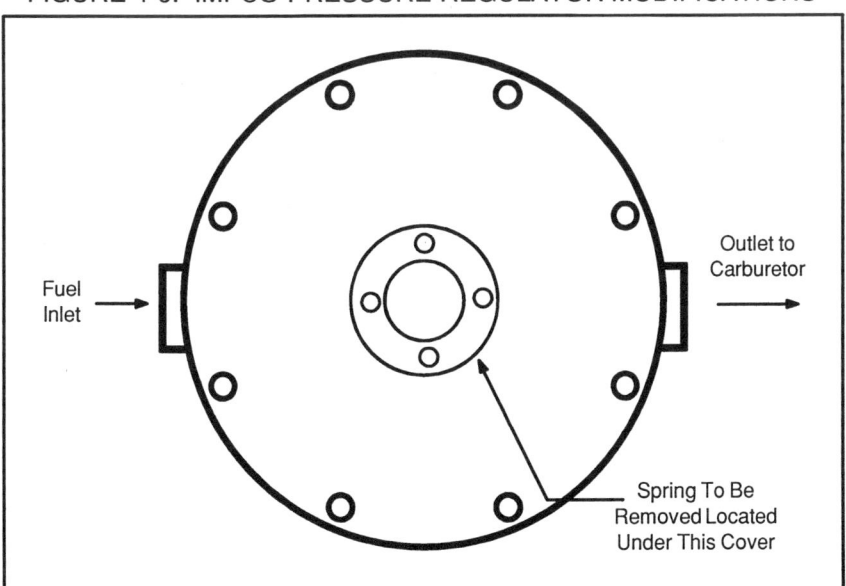

FIGURE 4-7: OXYGEN ANALYZER CALIBRATION
FOR READING AIR/FUEL RATIOS IN HYDROGEN ENGINES

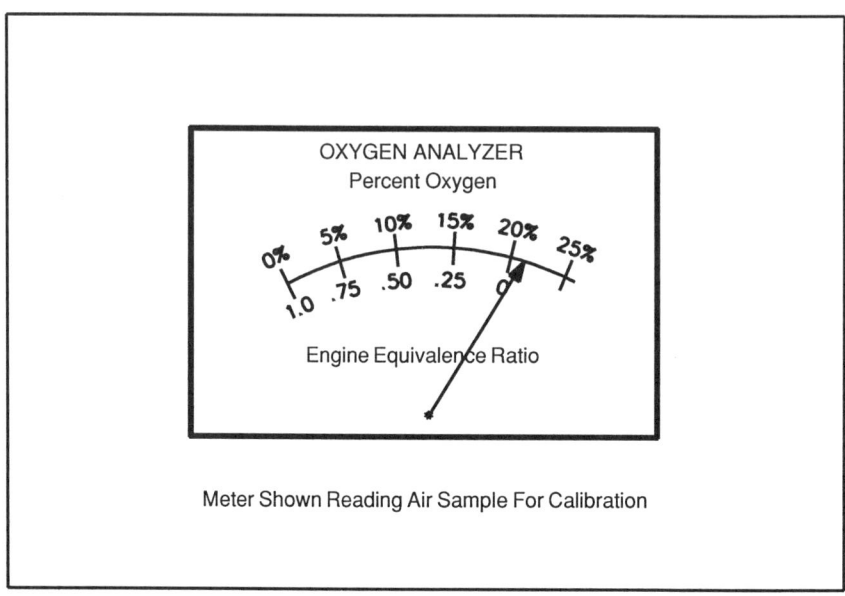

engine idling. With a spark advance setting of 10 degrees before top-dead-center, it is possible to idle a hydrogen engine at a mixture as lean as 0.15 to 0.2 equivalence ratio. This is 15 to 20 percent of stoichiometry, or the chemically correct mixture of fuel and air for proper combustion. (In simple terms, it is one volume of hydrogen for every two-and-one-half volumes of air.) In order to make the full power adjustment properly, a dynamometer is required to put the engine under a load while operating at maximum horsepower in the laboratory.

A very useful tool when testing or tuning hydrogen engines is an oxygen analyzer. By sampling the amount of oxygen remaining in the exhaust of the hydrogen engine, it is easily possible to determine the air/fuel mixture at which the engine is operating. Because of the chemical properties of hydrogen, this method is effective for mixtures ranging from 20 percent to 100 percent of stoichiometry. Such an instrument is easily calibrated using air which contains 20.99 percent oxygen. After calibrating the instrument, an exhaust reading of 10 percent oxygen would indicate an air/fuel mixture of 50 percent of stoichiometry. I usually affix a stick-on scale to the oxygen meter which is calibrated in equivalence ratios. For convenience, the conversion table from this calculation is provided as *Figure* 4-7. Since many of the less expensive oxygen analyzers are sensitive to heat and water, it is always a good idea to run the exhaust gas sample through a cooler and a water trap before passing it through the oxygen analyzer. This can be easily accomplished by making a coil of 1/4-inch metal tubing through which the exhaust gases pass. The coil can then be submerged in a container filled with ice water. The coil cools the exhaust gases and causes the water vapor to condense. The exhaust is then passed through a float-type water trap such as that manufactured by Armstrong Machine Works Inc. of Three Rivers, Michigan. The water trap allows the gases to pass through,

FIGURE 4-8: EXHAUST GAS MONITORING SYSTEM

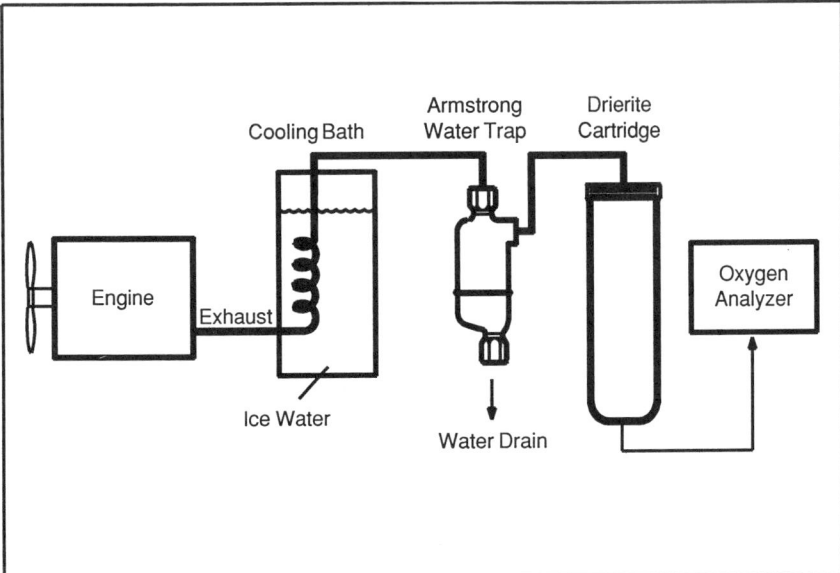

but the liquid water is trapped inside the device. When sufficient water is collected, a float lifts a drain valve in the bottom of the trap and the water is drained out. Since the oxygen is not completely dry at this point, it is also a good idea to pass the exhaust gas through a drying cylinder containing anhydrous calcium sulfate. This will remove the remaining moisture. Such a device marketed under the trade name of Drierite can be obtained from a chemical supply house. After use it can be rejuvenated in an oven and reused. A schematic diagram of a typical exhaust gas monitoring system is depicted in *Figure* 4-8.

Chapter 5

Advanced Engine Conversion Direct Cylinder Injection

CHAPTER 5

ADVANCED ENGINE CONVERSION DIRECT CYLINDER INJECTION

One of the major disadvantages of converting an engine to hydrogen is the resulting loss of 25 to 35 percent of the power from the engine. Interestingly, this loss of power is not an intrinsic characteristic of hydrogen engines. To the contrary, hydrogen, with its higher flame speed, actually has the potential of producing more power than the conventional gasoline fuels.

To understand this loss of power phenomenon, it is helpful to consider what actually takes place inside the internal combustion engine. During the intake stroke, a mixture of fuel and air is drawn into the combustion chamber. The carburetion system on the engine makes certain that the ratio of fuel to air is within the range for proper and complete combustion. In the case of gasoline, the fuel is in the form of tiny droplets of liquid. These droplets occupy such a small area that during the intake stroke they constitute less than one percent of the total volume of fuel and air drawn into the combustion chamber. The case with hydrogen is very different. Hydrogen is a very voluminous fuel. If the carburetor is set to mix hydrogen and air at 0.8 stoichiometry (which is a reasonable setting), then 25 percent of the volume of gas in the intake manifold is hydrogen and only 75 percent is air. In a cylinder having 1000 ml of displacement, 990 ml of air would be drawn into the chamber during the intake

Advanced Engine Conversion -- Direct Cylinder Injection 45

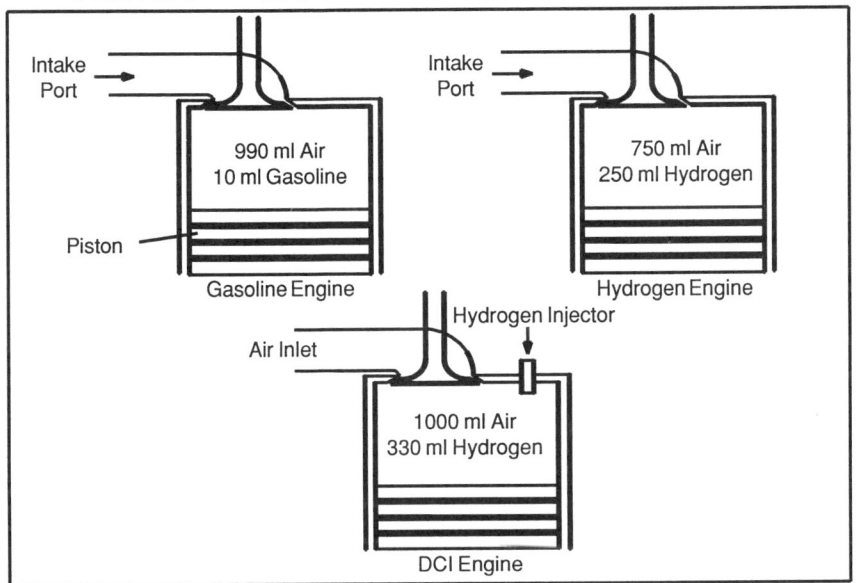

FIGURE 5-1: LOSS OF POWER IN HYDROGEN ENGINES

stroke if the engine was operated on gasoline. If the same engine was converted to hydrogen, the amount of air drawn into the cylinder during intake would be reduced to just 750 ml—a reduction of one fourth. The reason for this is that the combustion chamber would fill the rest of the way with the voluminous hydrogen. This phenomenon is depicted in *Figure* 5-1. Since one fourth of the air is displaced by the hydrogen, there is not as much oxygen available to react with the fuel and generate power. Since the amount of oxygen is reduced by approximately one fourth, the amount of power is proportionately reduced by a fourth.

The real culprit, then, which causes hydrogen engines to be less powerful than their gasoline counterparts is their inability to get a full charge of air into the engine during the intake cycle. The remedy to this problem is very simple. It is a technique known as direct cylinder injection (DCI).

In a DCI system there is no throttle limiting the amount of gases going into the intake port of the engine and no fuel pre-mixed with air. During the intake stroke the engine draws in a full charge of 1000 ml of air. After the chamber is filled with air, while the intake valve is closing, and at the beginning of and during the compression stroke, hydrogen is injected into the combustion chamber under pressure. By this method the loss of power problem is completely overcome. In fact, the pressurized hydrogen entering into the cylinder increases the pressure over that drawn in naturally by the engine, giving a sort of turbo-charged effect to the system, which results in not only equivalent but even enhanced power output as compared to operation on gasoline.

There are many benefits of such a system. Since there is no throttle on a DCI engine, the pumping losses or the loss in power due to low volumetric efficiency at part throttle settings is eliminated. In fact, this engine has the main efficiency advantage of a diesel engine. Furthermore, since the hydrogen is injected into the combustion chamber during the intake stroke, or, in some cases, at the end of the compression stroke, the gases in the chamber do not have an opportunity to mix completely. As a result, the charge is "stratified", resulting in a much slower flame speed and consequently lower peak combustion temperatures. In this way the tendency of the engine to detonate is greatly diminished and nitric oxide formation is reduced.

Another major advantage of DCI is its impact on backflash. When the intake valve opens in a DCI system, a cool charge of pure air is drawn into the combustion chamber. This cool air quenches any residual flame, hot catalytic particles, or any other source of ignition within the combustion chamber—all before the hydrogen fuel is introduced into the system. As a result, DCI systems do not have the backflash problem.

There are, however, some problems associated with the implementation of a direct cylinder injection system. The first problem is that the injection of the hydrogen gas must be carefully synchronized or controlled to be in perfect phase with the internal combustion engine. This is accomplished by placing a position sensor on the crankshaft which is used to electronically trigger a computer control system, or, alternatively, to mechanically actuate an injector system. The more difficult problem is to inject the large volume of hydrogen fuel required for each combustion cycle. To accomplish this requires an injector having a substantial orifice size. Injectors with a large enough orifice and which operate fast enough to keep up with the engine at high speeds are very hard to find and difficult to make.

When we first attempted to build a DCI system, we contracted with one of the nation's leading aerospace solenoid manufacturers to have a custom solenoid design developed for us. We provided to them the specifications which we required for response time, orifice size, and operating pressure. After many months of experimentation they gave up, explaining that to build an injector capable of having the performance characteristics we required would result in a very large device and consume a very substantial amount of electric power to open the large orifice. Obviously, if we wasted all of the power generated in the hydrogen engine just to actuate the electronic injectors, we certainly had not accomplished our initial objective.

In one prototype this problem was overcome by utilizing the diesel fuel injector pump from a diesel engine as the means of actuating the mechanical hydrogen injector valves. When it was time to inject fuel into the engine, the diesel fuel pump would pump high pressure oil into the injector which would then open the injector orifice, allowing the hydrogen to flow into the combustion chamber. The diesel fuel was then returned to the reservoir where it could be used over and over as the actuating fluid. The system performed satisfactorily, but was expensive and inflexible. What we really needed was an on board microcomputer to control the engine functions. A computer was desirable in this application because it could monitor engine conditions and compensate by optimizing hydrogen injection timing, duration, and spark timing.

The answer to the injector dilemma came from a colleague, a physicist by the name of Dr. Robert Ridge. Dr. Ridge noted that the intensity of a magnetic field increases exponentially as the inverse of the distance between two magnets. In other words, it is very difficult to pull a solenoid closed with an electromagnet, even if the magnetic materials are only a small distance apart. On the other hand, once two magnets are touching it is relatively easy and requires only a small amount of electric power to hold the magnets together even if a relatively strong force is pushing against them. Dr. Ridge saw an interesting way to take advantage of this phenomenon in terms of the internal combustion engine. In Dr. Ridge's design, the electromagnet in the solenoid was used to hold the valve closed, inhibiting the flow of hydrogen. It was very easy, then, to cut the power to the solenoid and allow the energy in the high pressure hydrogen to push the solenoid open, allowing the fuel to enter rapidly into the combustion chamber. This design then utilized the pressure caused by igniting the hydrogen inside the combustion chamber to push the solenoid valve back closed where it was latched by the electromagnet. We implemented this ingenious idea in a prototype device which was then installed in the prototype vehicle built for Peugeot.

The computer control system for the Peugeot prototype monitored throttle position, engine rpm, and oxygen content in the exhaust. Creating an engine performance look-up table by operating the engine manually on the dynamometer made it possible to teach the control computer the exact, optimum settings for injector timing and ignition timing under each operating condition. In actual practice we had some difficulty in isolating the prototype engine control computer from the high voltage engine ignition system. Eventually, it was necessary to use optoisolators on all of the lines connecting the control computer to the engine. The results, however, were most encouraging. We are now looking forward to applying this same technology to the higher compression ratio of a diesel-type engine.

Another DCI method utilizes an engine with two intake valves. Through one intake valve the incoming air charge is directed into the combustion chamber. The second valve is utilized to inject pure hydrogen directly into the combustion chamber.

I expect that hydrogen-fueled engines of the future will be based on DCI technology. Often these conversions will also employ water induction to minimize nitric oxide formation and to increase efficiency. Diesel engine conversions will be accomplished by fabricating injectors with a built-in spark plug capable of being retrofitted into the existing diesel fuel injector ports. Similar devices will be installed in gasoline-burning engines where the spark plug and the injector go into the combustion chamber through the existing spark plug opening. In this way, hydrogen engine conversions will be accomplished in the field easily and without any machining or drilling new holes into the combustion chamber. I expect that in production, such engine conversion kits will be available for popular engines at a cost of somewhere between $350 and $1,000 per conversion.

Chapter 6

Hydrogen Storage Compressed Gas and Liquid

CHAPTER 6

HYDROGEN STORAGE COMPRESSED GAS AND LIQUID

From the very beginning of my hydrogen career, the first question people ask concerning hydrogen research has always been, "Is hydrogen safe?" To the lay public, the term "hydrogen" is associated with the Hindenburg disaster or with the hydrogen bomb. In either case, the public perception is that hydrogen is a very dangerous substance. True, hydrogen is a combustible fuel capable of detonating in certain circumstances, but the inherent fear of hydrogen by the public (the so-called "Hindenburg syndrome") is really not warranted. If anything, the Hindenburg disaster is an example of the safety advantages of hydrogen. Imagine a giant floating gasoline fuel tank three football fields in length becoming ignited high above the ground with a host of passengers strapped to its "belly". The fact that two-thirds of these passengers survived the fiery disaster is a testament to the unique chemical properties of hydrogen. It is really one of the safest fuels known to man.

In the Hindenburg accident, the passengers very close to the giant hydrogen fireball were not seriously burned as would be the case with other fuels. This is due to the fact that no infrared energy is produced by hydrogen combustion. (The hydrogen flame is virtually invisible.) In fact, most of the passenger burns were actually caused by flames from the diesel fuel which was used on board to power the Hindenburg's engines. Yet the fear of hydrogen persists.

If we are going to transform our society to the use of a new, environmentally safe fuel, then we must do so in a way that greatly improves vehicular safety.

The foremost safety consideration in hydrogen transportation systems is the method of hydrogen storage. To date, numerous methods of storing hydrogen on a vehicle have been proposed. These include compressed gas, liquid, metal hydrides, solid hydrogen, micro-spheres, liquid hydrides, and finally, as a hydrocarbon which is reformed into hydrogen on board the vehicle. Most commercially viable hydrogen energy applications will employ one or more of these storage technologies, which we will discuss here in more detail (except for metal hydride storage which is the subject of the next chapter).

1. Solid Hydrogen

Interesting research is currently underway investigating the potential of forming solid hydrogen (slush). Although opinions vary regarding the potential of this form of hydrogen as a means of storage, the phenomenon is interesting and, if commercially developed, could be the lightest and most compact form of hydrogen storage available. At present the concept is a theoretical/laboratory curiosity and cannot be considered as a viable hydrogen storage alternative.

2. Microspheres

Researchers have investigated the possibility of storing hydrogen gas trapped inside tiny glass microspheres. This method takes advantage of hydrogen's ability to diffuse through hot glass. High pressure hydrogen diffuses through the thin glass walls of microspheres when the spheres are heated to high temperatures and the hydrogen is supplied at high temperature. The spheres are then cooled under pressure and the hydrogen gas is trapped inside, completing the charging procedure. When a vehicle begins to operate with hydrogen stored in microspheres, the spheres are heated utilizing exhaust gases. As the glass walls of the spheres become warm, hydrogen gas begins to diffuse through the glass and is released to fuel the vehicle.

Though somewhat functional and very novel, microsphere storage of hydrogen has serious drawbacks. The amount of hydrogen that can be stored in a vessel of a specified volume is not great. In other words, vessels containing microspheres for hydrogen storage are unreasonably voluminous. More important, studies have concluded that approximately 12 percent of the microspheres rupture during each charge/discharge cycle. This means that after recharging the vehicle just four times the storage capacity of such a vessel is cut in half. Though laboratory prototypes using this technology have been built and tested, the technology has never proven feasible.

3. Liquid Hydride

"Liquid hydride" is a term which I coined several years ago to describe liquids which have the ability to react with hydrogen, forming a secondary liquid which can then be dissociated back into the original liquid, releasing hydrogen to power a vehicle. Several liquids (usually organic compounds) have been considered as possibilities for such a system. In practice, the vehicle would be fueled with the form of the liquid hydride containing the surplus of hydrogen. As hydrogen is required for use in the vehicle's propulsion, the "hydrogenated" liquid would be broken down into the base liquid with gaseous hydrogen being evolved. The resulting liquid would be stored in a separate tank on board the vehicle.

Upon arriving at the refueling station, the broken-down liquid would be transferred to the station in exchange for the hydrogenated version. The depleted base liquid would then be re-hydrogenated at the station for use by the next vehicle.

Though the concept is novel and technologically viable, no base liquid has been identified which combines the properties of being safe, low cost, and able to store significant quantities of hydrogen in a small, lightweight container. Research performed to date has ruled out this option of hydrogen storage on board a vehicle.

4. Hydrocarbon Reformation

Many researchers have expressed enthusiasm about the potential of generating hydrogen on board the vehicle by a process known as hydrocarbon reformation. The most popular fuel being considered for hydrocarbon reformation is methanol. In this scenario the owner of a hydrogen car would refuel his vehicle by filling his tank with liquid methanol. The methanol would then pass through a catalytic reformer which would convert the methanol into hydrogen as required by the vehicle. Experts say methanol could easily be distributed within the existing hydrocarbon fuel infrastructure. (In other words, methanol could be distributed in existing liquid fuel tanker trucks, stored in existing underground storage tanks, and dispensed in existing liquid fuel pumps.)

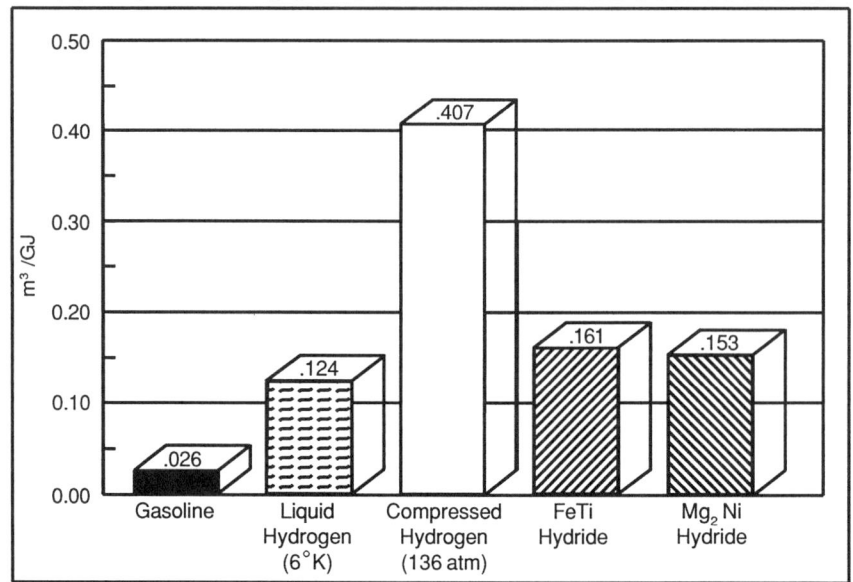

FIGURE 6-1: HYDROGEN STORAGE -- VOLUME COMPARISONS

One serious drawback to hydrogen reformation is that methanol and the other hydrocarbons considered for this application cannot be reformed into just hydrogen. In reality, along with the hydrogen, the reformation process results in a host of by-products including carbon dioxide, carbon monoxide, and many other compounds, depending on the source fuel impurities. All of a sudden, since you do not have pure hydrogen, you do not have all of the advantages associated with hydrogen such as improved efficiency, environmental cleanliness, etc. Though some people are very optimistic about the potential of on board hydrogen generation by hydrocarbon reformation, I personally remain skeptical and do not consider such projects to be real hydrogen energy implementations. (Furthermore, to date none of the attempts at utilizing this technology have really performed to expectations.)

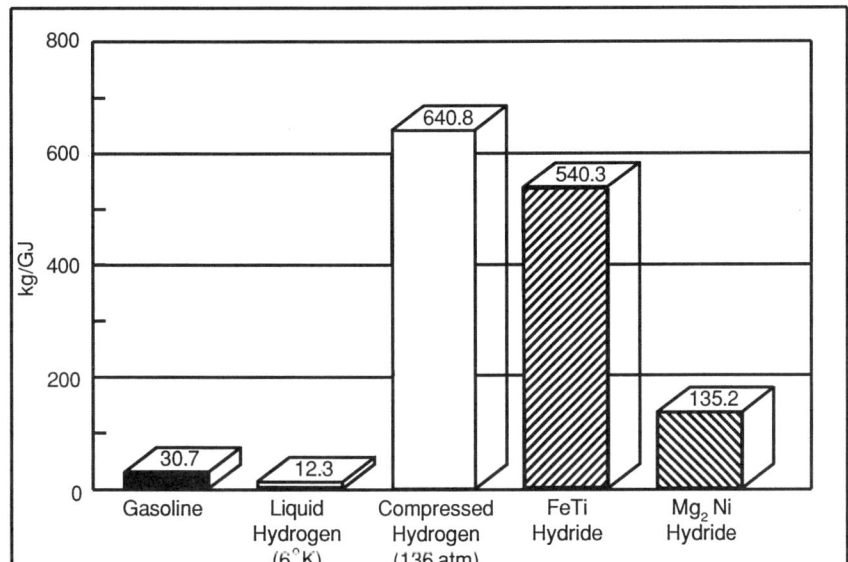

FIGURE 6-2: HYDROGEN STORAGE -- WEIGHT COMPARISONS

5. Compressed Gas

The best choices currently available for hydrogen storage on board vehicles are: compressed gas, liquid, and metal hydride technology. A comparison of the relative volume and weight of these storage technologies is presented in *Figures* 6-1 and 6-2.

Gaseous hydrogen is the most voluminous and the heaviest of the practical hydrogen storage technologies. It is also the most economical. For these reasons, compressed gas storage is the best method for stationary hydrogen storage installations.

6. Liquid

Liquid hydrogen is the preferred method of storing hydrogen for space and aircraft applications where the light weight is the most important consideration. As the lightest fuel known to man, hydrogen has tremendous potential in aerospace applications and, in fact, is the fuel of the space shuttle program. This method of storing hydrogen involves transforming the gas into its liquid form. This is accomplished by cooling hydrogen to 20.3° Kelvin (-422.9° Fahrenheit). The processes involved in cooling the gas to that very low temperature are energy intensive and require specialized equipment. Consequently, liquid hydrogen is considerably more expensive than gaseous hydrogen.

However, for aviation applications where extreme light weight is critical, liquid hydrogen is the preferred choice for fuel storage in spite of its higher cost. (Although hydrogen weighs exactly the same whether in liquid or gaseous form, the high strength containers needed to hold the high-pressure gas in its gaseous form are much heavier than the vessels used to store liquid hydrogen. Hence, stored liquid hydrogen is lighter than stored gaseous hydrogen.)

Liquid hydrogen is stored in "sophisticated thermos bottles". Any heat which leaks into the liquid hydrogen will cause a proportionate amount of the liquid to "flash" or vaporize into the gaseous form. For this reason, engineers have developed "super insulation", which is employed in cryogenic storage containers. (Cryogenic means having to do with extremely low temperatures.) The common method of making super insulated containers involves a lightweight, thin-walled inner vessel, which actually holds the liquid, surrounded by a jacket consisting of alternating layers of reflective mylar and an insulating layer of foam. The mylar and foam layers are encapsulated in a vacuum to further minimize the conduction of heat. The entire vessel is then enclosed in an outer vessel which makes possible the maintenance of the vacuum.

One of the difficult problems associated with the design of a cryogenic hydrogen vessel is the method of supporting the inner vessel. If it were to be supported inside the outer vessel by some type of a metallic bracket, that metallic bracket would conduct a significant quantity of heat into the inner vessel, causing the hydrogen to vaporize ("boil-off"). However, various clever technologies have been developed and are commercially in use to overcome this problem. In a well-designed cryogenic storage container, a boil-off rate of less than one-half-of-one percent is achieved. This means that the inner vessel is so well insulated that the amount of heat penetrating the insulation is only sufficient to vaporize one-half-of-one percent of the hydrogen every 24 hours. In other words, such a tank would be able to store liquid hydrogen for a maximum of 200 days. After that time, all of the liquid would have been transformed back into the gaseous state. Meanwhile, as the hydrogen changes back into the gaseous form it becomes voluminous and must be vented, or the pressure inside the liquid hydrogen tank begins to increase. In a liquid hydrogen vehicle, some hydrogen must be used almost every day. Alternatively, the vaporized hydrogen could be run through a re-liquefaction machine and put back into the tank, although this approach is more practical for large-scale applications.

The rate at which hydrogen is vaporized inside the cryogenic container is referred to as the "boil-off" rate. Typical boil-off rates range from one-half-of-one percent per day for very well-built tanks to 5 to 10 percent per day for less expensive vessels.

When a liquid hydrogen tank is filled for the first time or is re-filled after sitting for some period of time, the inner vessel must be cooled down to liquid hydrogen temperature. This is done by pumping liquid hydrogen into the inner vessel and allowing it to "flash-off". The hydrogen which is vaporized into gaseous form by the latent heat of the inner containment vessel is referred to as "flash-off" and must be dealt with when refueling liquid hydrogen containers. The amount of flash-off is usually equal to only 1 or 2 percent of the tank capacity. It is usually just vented into the atmosphere. Fortunately, flash-off only occurs when a liquid hydrogen tank is completely empty and has been empty long enough for the inner vessel to warm up to ambient temperatures. Nevertheless, a provision for handling the flash-off during recharge of liquid hydrogen containers must be taken into consideration when designing such systems.

To prevent hydrogen from vaporizing during the refueling process, it is transported in special vacuum-jacketed transfer lines. The vacuum jacket insulates the inner pipe, keeping the hydrogen cold. If liquid hydrogen is allowed to be transported in fuel lines which are not vacuum-jacketed, oxygen and nitrogen from the air which comes into contact with the line are cooled to below their liquefaction temperatures. As a result, liquid nitrogen and liquid oxygen drip off such a line. Liquefied oxygen can be dangerous. These cold temperatures can cause sever "cold burns" when brought into contact with the skin. Furthermore, the density of oxygen in liquid air can cause combustible materials to oxidize at more rapid rates than normal. For these reasons, and to prevent liquid hydrogen flash-off, liquid hydrogen is always transferred in vacuum-jacketed lines.

Although flash-off is a problem to contend with in liquid hydrogen storage systems, the tendency of liquid hydrogen to vaporize can also be useful. For example, it is possible to build a simple compressor or pressure builder for liquid hydrogen systems by installing a metal coil below the vessel. When a valve is opened, allowing liquid hydrogen to flow down into the coil, it becomes heated to gaseous hydrogen temperatures by the atmosphere. As a result, it expands or vaporizes. The coil is then connected to the top of the liquid hydrogen tank where the cold but voluminous hydrogen vapor causes the pressure within the vessel to increase. This method has been successfully utilized to pump the hydrogen from the liquid storage tank to the engine in our prototypes.

Our first liquid-hydrogen-fueled prototype automobile was a Chevrolet Monte Carlo. A cryogenic tank, manufactured for the storage of liquified natural gas by Beech Aerospace Corporation, was fitted into the trunk of the vehicle. A sophisticated liquid hydrogen control and vaporizer system was also provided by Beech Aerospace. The hydrogen was directed into the 5.7 liter (350 cubic inch) internal combustion engine with an Impco carburetion system as described earlier. Control of nitric oxide formation and the elimination of backfire in the prototype were achieved by the water induction technique.

Utilizing a full charge of liquid hydrogen, the vehicle had a range of approximately 240 kilometers (150 miles). It took about 5 minutes to refuel the vehicle if the cryogenic storage tank was cold. When the internal vessel was warm, (when the tank had been left empty for an extended period of time), the flash-off process extended the refueling to about 10 minutes.

Heat conducted through the vacuum-jacketed insulation into the tank caused hydrogen to evaporate at a rate of about 5 percent of the total tank capacity per day. Typically, the vehicle could be left without being driven for two to three days without the need to vent off excess gas pressure. In other words, the heat which leaked through the insulation into the tank caused some hydrogen to vaporize. The amount of hydrogen which vaporized caused the gas pressure inside the tank to increase. If the car was not operated every couple of days, the pressure inside the vessel would become too great, and a pressure relief valve would open, allowing the hydrogen to escape from the vessel. In the Monte Carlo prototype a catalytic oxidation module was fabricated to prevent the buildup of escaping hydrogen into the garage where the vehicle was stored. As the hydrogen escaping from the tank would pass across the surface of the catalyst, it would mix with air and be catalytically (flamelessly) reacted into water vapor.

We operated the Monte Carlo liquid hydrogen prototype very extensively, and, except for minor maintenance, it performed flawlessly. After completing an extensive set of scientific tests, I put the Monte Carlo into service as my personal automobile, driving it on errands, to and from work, and on short trips.

One Sunday, after returning from church, I punched the automatic garage door opener button to put the car in my home garage as was my custom. On this particular Sunday, however, as I drove up to the open garage door, I felt impressed to leave the vehicle outside on the front drive, which I did. Later that evening, when I was about to retire, I went outside for the purpose of putting the hydrogen vehicle away for safekeeping in the garage. Again, I felt impressed that I should leave it outside, which I did.

The next morning, about one hour before sunrise, I was awakened by the sound of a car horn outside. I jumped up, looked out the window, and saw the Monte Carlo prototype engulfed in flames. I ran outside in my pajamas and turned on the sprinkler with the intention of extinguishing the flame. Unfortunately, the sprinkler head had corroded onto the end of the hose, and in my haste I was unable to remove it. I turned the water on anyway, but the sprinkler only drenched me in my pajamas, giving me very little ability to direct water towards the burning vehicle.

The flames must have been burning for quite some time in order to melt the plastic steering wheel and cause the horn switch to short out, thereby creating the constant horn blast which had awakened me from my sleep. By now the noise had roused the whole neighborhood, and a dozen people surrounded the car, anxious to offer whatever assistance they could.

My first concern was for their safety. I knew that if the fire were to damage the liquid hydrogen vessel causing the vacuum insulated jacket to be compromised, the remaining liquid hydrogen would quickly be vaporized, rupturing the vessel and creating a potentially dangerous situation. I tried to communicate my concern to the "helpful" bystanders, instructing them to move away from the car. However, because of the noise of the horn, I could not be heard. In desperation, I drove them back with water from the sprinkler and then continued the task of putting out the fire myself. Some minutes later, drenched, in my pajamas, and now with a sizable audience, I was finally able to put the fire out. I then quickly confirmed that all of the hydrogen pressure had been released from the tank.

At that moment I heard sirens, signaling the approach of emergency vehicles from the fire department. Assured that the fire had been extinguished and that there was no remaining hydrogen in the cryogenic container, I opened my garage and, with the assistance of my wet neighbors, pushed the vehicle inside. My garage door was just closing as the fire and police emergency vehicles rounded the corner, followed by an envoy from the press. On the driveway were curious droppings of ash around the perimeter of where the vehicle had been standing. I was in my pajamas and completely drenched, along with my neighbors.

The first light of morning was just coming up over the east Provo, Utah mountain as the fire engines arrived. An excited journalist began to ask questions like, "What's going on?!"

Not feeling particularly inclined to discuss the matter with the press, I replied, "Nothing," and excused myself to go into the house and change my clothes.

Soon the excitement was over, the crowd had dispersed, and the press had given up hopes of finding a story. I, on the other hand, was in the deepest despair. After carefully examining the car, it became apparent to me that the cryogenic control system had malfunctioned, causing a hydrogen leak. The hydrogen had been ignited by the catalytic converter. This safety device had been the source of ignition, I supposed, and a small flame at the hydrogen storage vessel had spread to the combustible materials used in constructing the car. As I gazed upon the destroyed vehicle, I was convinced that liquid hydrogen was not an appropriate fuel for consumer vehicles. Its use would be limited to aviation or aerospace applications where the intensive maintenance associated with these applications would provide an adequate environment for safely operating the sophisticated equipment required in liquid hydrogen systems . . . but for automobiles, it was my impression that liquid hydrogen would never do.

For the next two days I was preoccupied in serious reflection and soul-searching. So much that was good about hydrogen this world needed, but it had to be made safe. There had to be a better way.

Chapter 7

Hydrogen Stored As Metal Hydride
The Safest Fuel on Earth

CHAPTER 7

HYDROGEN STORED AS METAL HYDRIDE THE SAFEST FUEL ON EARTH

About this time Reilly and Wiswall, researchers at the Brookhaven National Laboratory, proposed the possibility of storing hydrogen on board a vehicle by reacting it into weakly-bonded inter-metallic compounds referred to as "metal hydrides". These researchers had conducted laboratory experiments which indicated that such a fuel storage system would be feasible. However, no one knew how to build a full-size system. My colleagues and I had attempted to build what would have been the world's first automotive metal hydride storage for the Monte Carlo prototype. The results of that effort were most disappointing. Could the numerous problems associated with a metal hydride storage system be solved? Was this extremely safe method for storing hydrogen on board a vehicle really viable? I resolved to find out.

A metal hydride is a powdered material made up of a combination of carefully chosen metals. When everything is exactly right and after removing the oxide coating layer which forms on the surface of most metals, it is possible to create a metallic alloy which has the property of reacting with gaseous hydrogen. The resultant powdery material has neither the properties of the metal nor the hydrogen. In some ways it is similar to table salt. Salt is a combination of the metal, sodium and the gas, chlorine. Sodium is a metal which is so reactive that it will burst into flame and even explode when brought into contact with water. Chlorine, on the other hand, is a poisonous gas. When the two are reacted together, they become table salt, a substance with completely different properties from the original elements. In a similar way, when hydrogen gas is reacted with certain metal alloys, a hydride is formed. If the metal alloy is carefully selected, it is possible to make a metal hydride which is reversible. In other words, hydrogen gas can be reacted with the metal alloy to form the hydride and then, by gently heating the hydride, it can be dissociated again into hydrogen gas and metal powder. In a way, the hydride is like a sponge which soaks up very large quantities of hydrogen gas, converting it into a compact and safe form for storage, but then releases the hydrogen as needed by the vehicle.

In order to make an automotive hydride tank store hydrogen, it was necessary to learn a lot about the dynamic chemical properties of the substances involved. The chemical properties of reaction with hydrogen vary greatly from metal to metal. This becomes especially complex when two or more different metals are mixed together into an alloy of varying percentages, each with different hydriding properties.

To get a handle on the chemical properties of the various metal alloys, we built a data collection device which we called the auto-hydrider. This machine would test the chemical hydriding properties for a specific alloy which was placed into the machine. It would do these tests very quickly and dynamically, thereby giving us the kind of data necessary to do the engineering of real world metal hydride storage systems. Not being able to afford one of the expensive computers on the market in 1974 to control the auto-hydrider, I investigated the potential of utilizing a microprocessor. Intel Corporation had just recently announced the 8080A microprocessor, which turned out to be exactly the device we were looking for. This early machine led to my involvement in manufacturing one of the world's first microcomputer systems.

Over 200 different alloys were characterized and studied to determine which would be most applicable to an automotive storage system. Several difficult problems needed to be addressed. Some alloys were extremely heavy, storing only small quantities of hydrogen. They would be too heavy for use in an automobile. Others were extremely expensive and therefore impractical. A third category of metal hydride alloys were dangerous. For example, mischmetal hydride is pyrophoric, meaning that it spontaneously ignites if brought into contact with air.

Finally, the iron-titanium family of metal hydrides was chosen for the first prototype vehicle. The powdered alloy was placed into stainless steel tubes which were manifolded together inside a jacket. During operation of the vehicle, exhaust gases from the engine would flow over the outside of these stainless steel tubes, providing the heat necessary to dissociate the hydride. The researchers at Brookhaven National Lab had suggested that a small gas reservoir would be needed to provide the hydrogen necessary to start the engine and to warm up the metal hydride storage container. We learned, however, that such a reservoir was not required. Enough hydrogen was available from even a very cold hydride vessel to get the system started.

The world's first successful vehicle hydride storage vessel was installed in a Pontiac Grand Ville. This time the system worked to perfection. To fill the vessel a hydrogen gas line was connected to the metal hydrogen storage tank. The procedure was very much like inflating a tire on a car. The gaseous hydrogen, which was at 18 atmospheres pressure (250 psig), flowed into the tank where it reacted with the metal hydride powder. When the hydrogen and the alloy combined, heat was liberated. The hydride storage tank could be charged as quickly as this heat could be dissipated. In the Pontiac Grand Ville the heat was discharged by blowing air through the tank. Utilizing this method, it took about 4 hours to recharge the vehicle with hydrogen. Refinements to the system, employed in later prototypes, reduced the recharge time to less than 15 minutes.

At last we had a way to store hydrogen safely. One of the first companies to become interested in the new hydride technology was Daimler-Benz (Mercedes). Engineers from Germany visited our Provo, Utah, research laboratories to learn about the new metal hydride technology and our method of converting the internal combustion engine to hydrogen fuel. When they returned for their second visit, it seemed that they were convinced that we had not really accomplished what we had claimed. The German engineers crawled underneath the vehicle, tracing the fuel lines from point to point in an attempt to determine where the fuel was "really" coming from. Finally they conceded that it was all "real" and went back to Germany where they built a "similar" prototype of their own. (It seems as though my career in hydrogen has been a series of "technological breakthroughs". The definition of a "technological breakthrough" is something that, when you first do it, no one believes you really did it; then, when it has become the industry standard, no one remembers who did it first.)

Of course, there were still a lot of problems with early hydride systems. After recharging the vehicle hundreds of times, we began to encounter a problem. Each time the metal hydride storage tank was recharged, the tiny powder particles would break down, becoming smaller and smaller. This effect, called decrepitation, did not impact the hydride's ability to store hydrogen, but eventually the particles became so small that they began to pass through the 5-micron sintered metal filter which we had installed to keep the hydride inside the tank. Later we learned that by adding a tiny amount of a third metal (manganese), the decrepitation process would discontinue when the particles reached a size of about 10 microns, thereby alleviating this problem.

Many prototypes and experiments were built, expanding our knowledge of metal hydride storage systems. Several patents relating to this technology were applied for and granted. Unfortunately, there was one major drawback of the metal hydride systems which we were unable to solve. Metal hydride storage systems are heavy. The auto makers told us that the fuel tank could not weigh more than 225 kilograms (500 pounds). With a tank of that weight, the vehicle range was limited to about 240 kilometers (150 miles).

Realizing the tremendous potential of metal hydride storage systems, researchers around the world have begun looking for lighter hydride alloys that would reduce the weight of a metal hydride storage system. Though millions of dollars have been spent looking, no one has succeeded in achieving an automotive metal hydride storage container very much lighter than our early prototypes. (Only now has this weight problem been resolved—not with better hydrides, but with more efficient fuel utilization. For more information, see Chapter 11 on fuel cells.)

Metal hydride storage containers are now manufactured out of aluminum at a substantial reduction in price and weight compared to the earlier stainless steel prototypes. Instead of running the exhaust through the hydride container, cooling water from an internal combustion engine or exhaust gases from a fuel cell provide the source of heat required for hydride dissociation. Numerous prototypes have been built, each one advancing the state of the art in some way. More concerning the vehicle prototypes we have built and the hydrogen storage methods utilized is discussed in Chapter 8.

As part of a study to determine the safety of the iron-titanium-manganese metal hydride storage systems, we participated in tests conducted by the U.S. Army. The tests consisted of firing armor-piercing incendiary bullets at fully-charged hydride containers. These special bullets simulate the very worst possible condition which could result from a serious collision. The dramatic footage taken of these tests graphically demonstrated that the metal hydride vessels do not explode. As the vessel was ruptured by the impact, the melting flame of the incendiary bullet ignited the small quantity of hydrogen gas which is always present, filling the void spaces in between the hydride particles. After the initial impact, a small, pilot-light-sized flame burned for several hours at the hole where the bullet had entered the vessel, as gaseous hydrogen was slowly released from the hydride material. These systems are so safe that we have actually been able to weld metal hydride storage vessels charged with hydrogen. Just imagine what would happen if someone were to attempt to weld a fuel tank filled with gasoline!

Another important attribute of metal hydride storage systems is their longevity. They do not "wear out". Metal hydride storage containers have been tested by charging and discharging them over and over in thousands of cycles, equivalent to millions of miles of driving. During the first 100 cycles, the hydride actually improves with each cycle. After the first 100 cycles, the improvement phenomenon is no longer observed, but there is no degradation in the performance of the tank even after thousands of cycles. For this reason, used metal hydride storage vessels will be salvaged from old vehicles, reactivated, and used again.

Although metal hydride storage systems don't wear out, hydrides can be poisoned by impurities in the hydrogen, including such things as carbon monoxide, oxygen, or even water vapor. However, because of the unique chemical properties of hydrogen, it is very easy to remove these impurities from a hydrogen gas stream even on a small scale, such as in electrolysis. Electrolysis units, which we have been building for the past 10 years, produce hydrogen directly out of the cell at pressures adequate for recharging hydride vessels. Although the electrolyzers will work at pressures of up to 40 atmospheres (600 psig), only 17 atmospheres (250 psig) is necessary to recharge a metal hydride storage vessel. Simple, yet reliable methods of purifying the hydrogen to the hydride grade have been successfully employed in the electrolysis process. These methods include de-oxo catalysts and pressure swing adsorption technologies employing molecular sieves.

Metal hydrides can also be employed in a variety of other applications ranging from hydrogen purification to hydrogen compression. In one of my patents (U.S. Patent #4,108,605H2), metal hydrides are used to purify hydrogen. It is also possible to charge a metal hydride vessel with low pressure hydrogen, and then, by heating the metal hydride, releasing hydrogen at pressures in excess of 700 atmospheres (10,000 pounds per square inch). This is accomplished without pistons, compressors, or any moving parts.

Chemical properties of three common metal hydrides are presented in *Figure*s 7-1, 7-2, and 7-3. The data is presented by depicting the

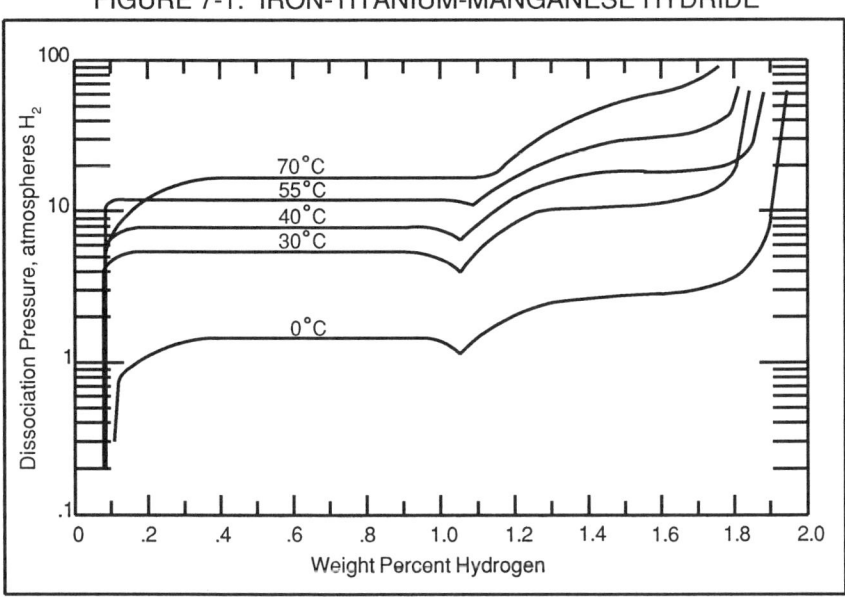

FIGURE 7-1: IRON-TITANIUM-MANGANESE HYDRIDE

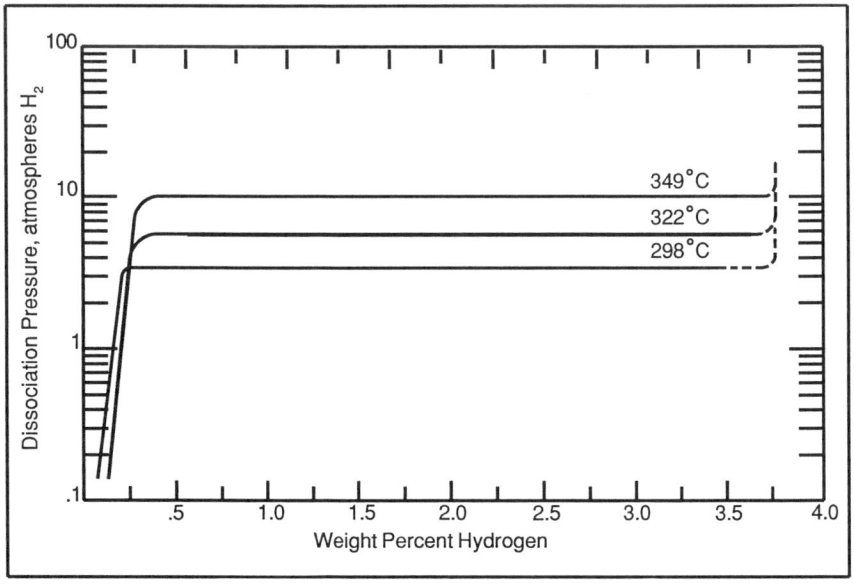

FIGURE 7-2: MAGNESIUM-NICKEL HYDRIDE

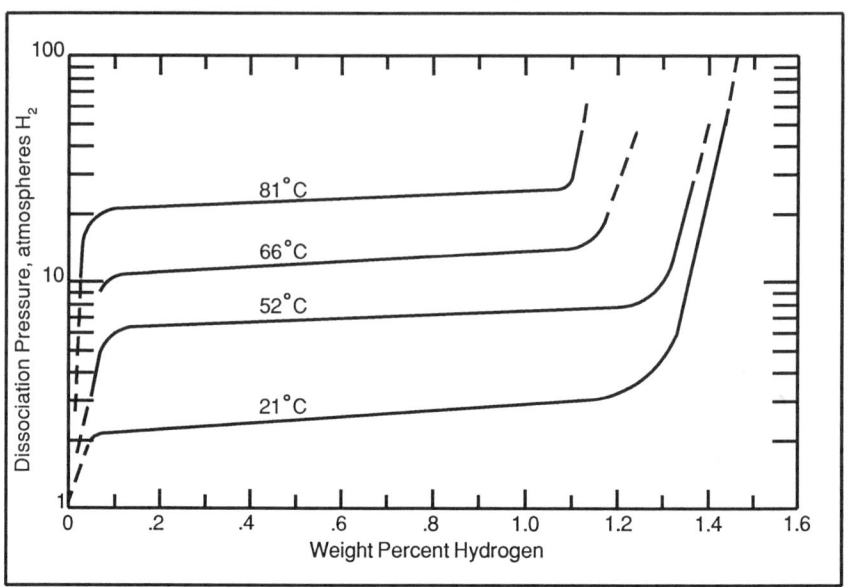

FIGURE 7-3: LANTHANUM-NICKEL HYDRIDE

pressure of hydrogen in a metal hydride storage container, at a specified temperature, as a function of the weight percent hydrogen. In a hydride storage container, the higher the temperature, the higher the pressure. Typical hydriding alloys have a pressure plateau, or in some cases, several pressure plateaus. As the tank is charged, the pressure rises rapidly until it reaches the plateau pressure for that particular temperature. Then, as additional hydrogen is added to the container, the hydrogen begins to be absorbed by the hydride in the vessel, and the pressure does not increase. This phenomenon continues until the reactive sites within the metal alloy are completely occupied. Then additional hydrogen causes the pressure to increase again, indicating that the tank's storage capacity is saturated.

The quantity of hydrogen stored in a metal hydride is called weight percent. This terminology refers to the percent of the weight of a vessel which is made up by hydrogen. For example, if a metal hydride has a weight percent of 2, then 100 kilograms (220 pounds) of that hydride could store up to 2 kilograms (4.4 pounds) of hydrogen. Having only a 2 weight percent sounds like a very low hydrogen storage capacity, but when you remember that hydrogen is an extremely light fuel, you see that it is better than it sounds.

The iron-titanium family of hydrides store hydrogen with weight percents ranging between 1.6 and 2.3. Other alloys have much higher weight percents. For example, magnesium hydrides have hydrogen weight percents as high as 7.6. Unfortunately, these alloys require very high temperatures to release the hydrogen, making them impractical for most automotive storage applications. The energy required to dissociate various metal hydride alloys is shown in *Figure* 7-4.

FIGURE 7-4: HEAT OF FORMATION OF METAL HYDRIDES

Material	Hydrogen Weight Percent	Heat of Formation
FeTi (trace Mn)	1.9	-3.4 kcal/gram H_2
Mg_2Ni	3.6	-7.7 kcal/gram H_2
Mg	7.6	-9.3 kcal/gram H_2
$LaNi_5$	1.5	-3.6 kcal/gram H_2

Hoffman and Reilly, BNL, 1976.

Metal hydrides provide a safe, compact, and affordable method of storing hydrogen on board a vehicle. Iron-titanium-manganese hydride stores hydrogen at approximately one-ninth the weight of the lead acid batteries used in current electric vehicles.

Chapter 8

Hydrogen Prototype Vehicles

CHAPTER 8

HYDROGEN PROTOTYPE VEHICLES

During my twenty-five year quest to launch the hydrogen energy economy, I have enjoyed the opportunity to work on a large diversity of research projects, each one designed to accomplish some specific goal or purpose. Included in these projects have been the conversion of 24 engines and propulsion systems, 18 of which have actually been installed and operated in vehicles. Following, then, is a chronicle of these hydrogen conversion projects:

Briggs and Stratton Lawn Mower Engine - 1964, 1965

My first engine conversion was the Briggs and Stratton engine converted in high school for the science fair. The details of this conversion have been described earlier. The conversion employed a simple variable equivalence ratio carburetor. The engine had a 6.5-to-1 compression ratio and used the stock ignition and timing curves. The prototype established that an SI (spark ignition) internal combustion engine could be successfully operated on hydrogen.

Model A Ford - 1965: World's First Hydrogen Car

The second conversion was a four cylinder, 1931 Model A Ford engine, converted in the vehicle to demonstrate that hydrogen could operate a vehicle. This was the world's first hydrogen car. Earlier vehicles had been operated on hydrogen-CO gas mixtures (town gas), but such conversions are hydrocarbon vehicles, not hydrogen. Several researchers had operated engines on hydrogen by that time, but there are no reports to indicate that any of those engines were ever installed into an automobile.

On the Model A prototype it was discovered that nitric oxide pollution was a potentially serious problem of hydrogen engines.

Ford V-8 Engine

After the science fair in high school, I enrolled at Brigham Young University with a composite major in Chemistry, Physics, Electrical Engineering, Chemical Engineering, and Mechanical Engineering. I also took courses in accounting, advertising, and communications with the goal of someday establishing a company to pursue my hydrogen energy dreams. During the five-year period before I received my Bachelor of Science degree, I continued to research hydrogen engine conversion technology with a grant from the Ford Motor Company.

The Ford Motor grant came about during the latter part of my freshman year after I wrote an unsolicited proposal and sent it to Ford. They responded with a letter of acceptance and a contract that required the signature of an officer of the university. Since I did not have a faculty advisor at the time, I took the contract to the Administration Building where I met the secretary to the Academic Vice President. Explaining what I needed, she asked me to leave the material with her, which I did. The next day I returned to find that the application was signed and ready for me to send back to Ford.

The contract stipulated that the university provide me with laboratory facilities in which to conduct the research. Accordingly, I made an appointment with the Vice President to ask him where I should set up the research. You should have seen the look on his face when I explained that I was an undergraduate—a freshman! He made good on his commitment to Ford, however, and laboratory space was provided in the Eyring Science Center.

My goal was to eliminate the nitric oxide from the hydrogen engine. To do this, I started by modeling the chemistry of the engine combustion chamber using a program developed by the Naval Ordinance Testing Laboratory. After trying exhaust gas recirculation and dozens of other options, I experimented on the computer with injecting water droplets with the fuel. According to the computer, the water injection completely eliminated the nitric oxide formation. To determine the effectiveness of this approach, I proceeded to convert a 1972 Ford V-8 engine which was donated for this purpose by the motor company. At first it was very difficult to get the engine, which had a hemispherical combustion chamber, to even run on hydrogen. Unlike the Model A, it had a severe tendency to backfire. Finally, when I began to apply the water injection, I was successful in getting the engine to run without backfiring.

The gas chromatograph equipment available to me to measure the nitric oxide in the exhaust revealed a reduction in the pollutant coming from the engine while utilizing the water induction, but it lacked the resolution to tell how good we were really doing. I needed someone with better instruments to test the car.

Mazda RX-2 - 1972

It was at this time that I heard about the "clean air race" competition between universities, which was just four weeks away. The Urban Vehicle Design Competition (UVDC) was the second running of the race, and this year vehicles would be tested at the General Motors Proving Grounds in Ann Arbor, Michigan. If I could get there, I would have a chance to have my hydrogen engine accurately tested.

I obtained, at no cost, the use of a new Mazda RX-2 with a Wankel engine from Ken Garth Mazda in Salt Lake City. After converting the engine to hydrogen with a water injection system, I learned by lab tests that the converted vehicle produced unexpected quantities of carbon monoxide and hydrocarbons. Then, with just ten days to go before the contest, I learned that the Mazda Wankel engine injects oil into the combustion chamber to lubricate the apex seal. The oil and my water injection system were not compatible.

Volkswagen Superbeetle - 1972

Not willing to lose an opportunity to have the water injection technology tested, I asked Peterson Motors of Provo, Utah for permission to convert a new Superbeetle to hydrogen for the contest. Chuck, the owner, asked if it would "hurt" the car. I promised to be very careful. The Volkswagen conversion was finished early the morning it had to leave for the contest. There was no time for a single emissions test before we left.

Although the testing of the Superbeetle has previously been discussed, some additional detail is interesting. The night before our test in Ann Arbor, the vehicle was taken into the laboratory for a "cool down" period. Going into the lab, the vehicle drove flawlessly. The next morning, when it was time to begin the test, I could not get the engine to even fire. In a near panic, I opened the rear engine compartment, looking for the problem. Everything looked fine. Still it would not start. By now the judges were calling our name. It was our turn. Each team got only one chance. After a brief glance heavenward, I inspected the engine compartment one last time. Then I saw it. Someone had removed the throttle return spring from the carburetor. No wonder it would not start. The engine was getting pure hydrogen. I borrowed a rubber band from the front office, tied a knot to make it fit, and installed it on the engine. Immediately the engine started.

On the dynamometer test stand, everything started out all right. I was still a little shaken by the "close call" but optimistic. Then, in the middle of the test, the engineer in the room behind the glass stopped the test. The same thing had happened earlier during the test of the ammonia car from the University of Tennessee. That car had emitted so much nitric oxide that its concentration could not be measured by the instruments, so the test was "scrubbed". Knowing the potential for producing nitric oxide in a hydrogen engine and realizing that we had not performed any testing of the car before the contest, I feared the worst. Later I learned that the engineers had to recalibrate their instruments on a lower scale to read the extremely low emissions from our car.

The Volkswagen prototype demonstrated that it is possible to operate a hydrogen engine without producing nitric oxide.

Briggs and Stratton II - 1973

Fresh out of school, with a wife, one child, and no money, I needed a place to start my research.

The Utah State Mental Hospital in east Provo had a vacant building of about the right size. The problem was to get permission. I tried to schedule a meeting with the hospital director but to no avail, so I arranged a meeting with Utah State Governor Calvin Rampton in Salt Lake City. At the meeting I showed the Governor the awards from the science fair, the UVDC, and various news clippings, explaining that I was now ready to start work on the project full time. "I notice there is a building vacant at the mental hospital near the university—I wonder if it would be OK with you, if, of course, the hospital director agrees, for me to use that building for my experiments," I queried. The Governor explained that he did not care. The matter was entirely up to the hospital director. Now, how to convince the director?

Again I had a hard time getting a meeting with the hospital director. When the receptionist asked what it was concerning, and could someone else help me, I was desperate. "I just want to discuss a meeting I had last week with Governor Rampton," I replied. I was sent right in.

The director seemed very busy. "What is this about a meeting with the Governor?"

"I met with the Governor to see if it would be possible for me to use the empty building outside to do experiments on hydrogen engines. The Governor said it was OK with him, but that the final decision was up to you," I explained.

"He said it was OK with him? The decision is up to me? What is that supposed to mean? Well, OK, but if you cause any problems you will have to go."

Now I had a rent-free building to begin my research, and, to many, I was at last where I belonged with my hydrogen dreams—on the grounds of the State Mental Hospital.

In the new lab we set up a test stand with a new Briggs and Stratton engine, and with a grant (our first) from the Charles Kettering Foundation, we were able to create a detailed engine map comparing performance on iso-octane and hydrogen. The data was published in two papers and gave us a launch.

Ford Falcon - 1973

In 1973, I converted my Ford Falcon in an effort to study the detrimental effects of water injection in the hydrogen engine. We found that if the water droplets were not properly atomized, or if too much water was injected into the engine, there was a problem of water ending up in the oil pan. The water, if excessive, would froth up into a white foam and overflow onto the floor. We learned that the best way to control this problem was to prevent it in the first place by limiting the amount of water injected and making certain that it was properly atomized.

Monte Carlo - 1973: First Liquid Hydrogen Automobile

A cryogenic storage vessel fabricated by Beech Aerospace was installed into the trunk of this prototype. The tank stored 3.75

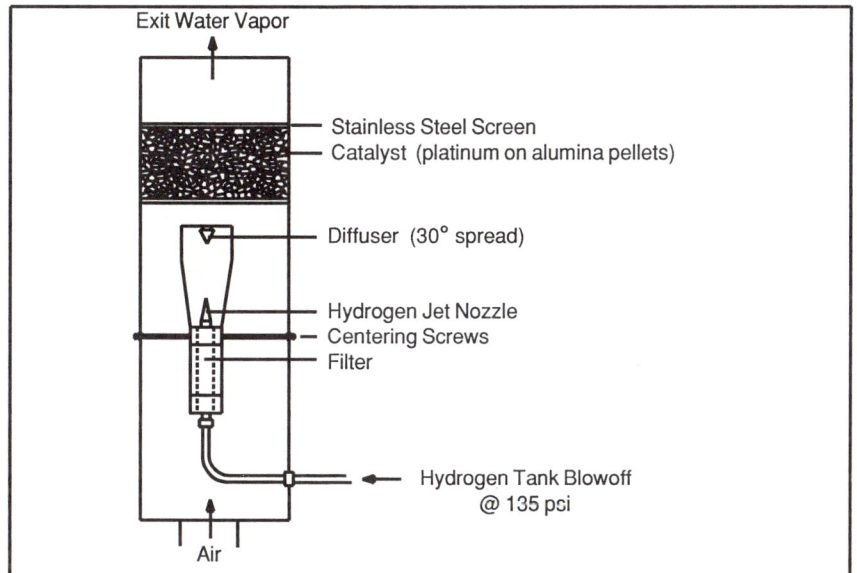

FIGURE 8-1: CATALYTIC REACTOR TO HANDLE LIQUID BOIL OFF

kilograms (8.3 pounds) of liquid hydrogen, enough to give the vehicle a 160 kilometer range (100 miles). The tank had a lock-up time (from 1 to 5 atmospheres) of two days if the tank was fully charged. After two days, the vessel would begin to vent hydrogen through a catalytic reactor to prevent dangerous quantities of hydrogen from accumulating external to the vehicle. *Figure* 8-1 is a schematic drawing of the reactor.

In addition to the liquid system installed in this prototype, we also made our first attempt at a vehicular metal hydride storage system. It was thought that the hydride would be able to capture hydrogen boil-off rather than wasting the fuel.

Unfortunately, the Monte Carlo metal hydride system did not operate successfully. Part of the problem was the quality of that

86   The Hydrogen World View

early batch of "homemade" hydriding alloy. The biggest problem, however, was that the metal tubes into which the alloy was placed could not hold hydrogen. The hydrogen leakage was so excessive that the vessel could not even be tested properly. Hydrides would have to wait for another prototype, but the liquid system performed to expectation, until, after about 12 months of driving, when the liquid tank fill solenoid malfunctioned, and the vehicle was destroyed by fire.

Pontiac Prototype - 1974: World's First Hydride Vehicle

Our second attempt to build an automotive metal hydride storage vessel was much more successful. This tank was made of welded stainless steel tubes. The tube bundle was covered by an outer shell, and engine exhaust was allowed to flow around the outside of the

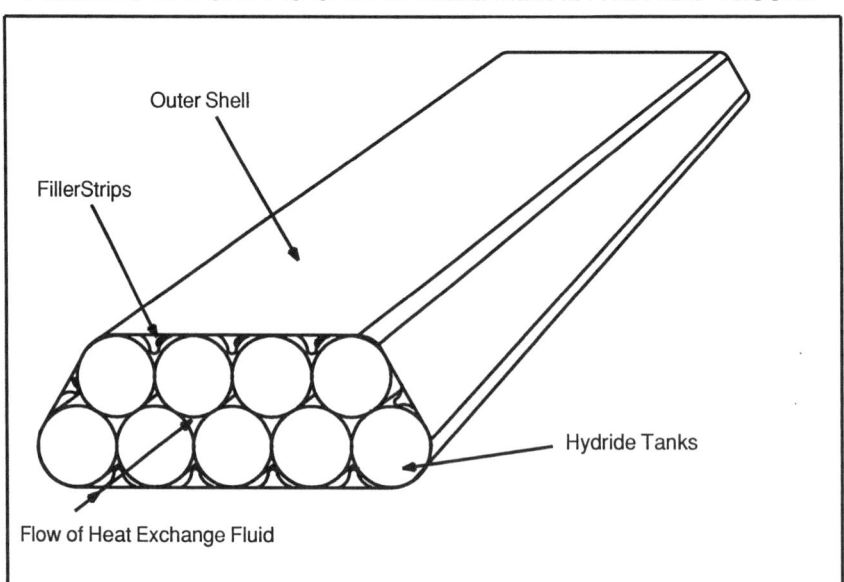

FIGURE 8-2: PONTIAC GRAND VILLE METAL HYDRIDE VESSEL

FIGURE 8-3: GRAND VILLE PROTOTYPE HYDRIDE SPECIFICATIONS

Mass of Hydride	198 kg (436 lbs)
Mass of Pressure Vessel	136 kg (299 lbs)
Total Mass	334 kg (735 lbs)
Hydride Material	Iron-Titanium
Recharge Pressure	34 atm (500 psi)
Recharge Temperature	16°C (60°F)
Hydrogen Capacity	2.4 kg (5.23 lbs)
Weight Percent Hydrogen	1.2%
Usable Hydrogen at 80 km/hr	1.8 kg (3.9 lbs)
Performance	25 km/kg (7.0 mi/lb)
Range	44 km (27.3 mi)
Top Speed	145 km/hr (90 mph)

tubes to provide the heat necessary to dissociate the hydride. The tank layout is shown in *Figure* 8-2. The technical specifications of the Grand Ville hydride system are presented in *Figure* 8-3.

Although the maximum range of the prototype was 60 kilometers (37 miles), the hydride could not keep up with the engine at highway speeds. At 80 kilometers per hour (50 mph) the vehicle range was reduced to 44 kilometers (27.3 miles). Though the range was disappointing, the fact that this vehicle was actually operating from hydrogen stored in the powder form was very exciting. Not only could the vehicle be recharged with moderate-pressure gaseous hydrogen which is considerably less expensive than liquid, it was also very safe. The hydride exhaust system used in the Grand Ville prototype is depicted in *Figure* 8-4.

FIGURE 8-4: EXHAUST HYDRIDE SYSTEM INSTALLED IN THE GRAND VILLE

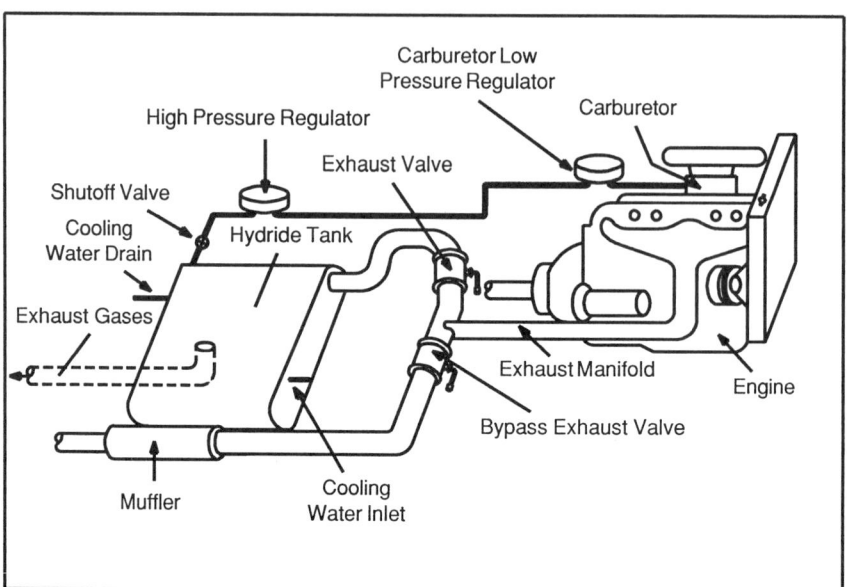

Winnebago Motor Home - 1974

An Indian model Winnebago motor home was modified to operate on hydrogen. The fuel was stored in a 150-liter liquid hydrogen dewar built by the Linde Division of Union Carbide. The liquid hydrogen was vaporized in a specially designed heat exchanger utilizing engine heat transported through the cooling water system. The Dodge 7.2-liter (440 cubic inch), eight-cylinder engine was converted to hydrogen with a modified Impco carburetor and a water injection system. The Onan generator, which provided 110-volt power to the vehicle, was converted using the variable equivalence ratio carburetor similar to the original Briggs and Stratton engine.

The propane appliances were also converted. They consisted of a range, a water heater, a forced-air furnace, and a gas refrigerator. All of the propane appliance conversions were made by eliminating primary air and by making minor modifications to the burners. The vehicle operated extensively for four years with very good results.

Citicar - 1975

This electric vehicle was converted to hydrogen to compare the range of operation with lead acid batteries and with a metal hydride hydrogen system.

The electric motor was removed and replaced with a Kohler, two-cylinder gasoline engine which had been converted to hydrogen. The lead acid batteries were removed from their "under the seat" compartment, and an Fe-Ti hydride vessel of the same weight was fabricated to take their place. Before conversion, on a new set of batteries the car could travel 64 kilometers (40 miles) before refueling. After installation of the hydrogen system, the range was tripled to 190 kilometers (120 miles). As the second successful metal hydride prototype (the first with good range), the car received wide interest. It was featured at the Brookhaven National Laboratory Energy Fair in 1976, and was also the first hydrogen car to be demonstrated on Capital Hill in Washington, D.C.

Provo-Orem Bus - 1975: First Hydrogen/Hydride Transit Vehicle

The first really serious metal hydride vehicle to go into commercial service was this Winnebago 21-passenger bus which was put into public transit service on the route connecting the cities of Provo and Orem, Utah.

The bus was powered by the same type of Dodge engine we had converted for the Winnebago motor home earlier. The water injection system was improved by installing a means of increasing the water induction mass ratio when operating at or near full power settings. Engine performance data as a function of the equivalence ratio and water injection mass ratio are presented in *Figures* 8-5 and 8-6.

The bus route was 21 kilometers (13 miles) per one-hour lap of urban driving. A normal shift for a driver was 4 laps, or 4 hours of driving between breaks. To provide the bus with sufficient range to complete four laps, two Fe-Ti hydride vessels were fabricated, each having a total mass of 714 kilograms (1575 pounds) for a total

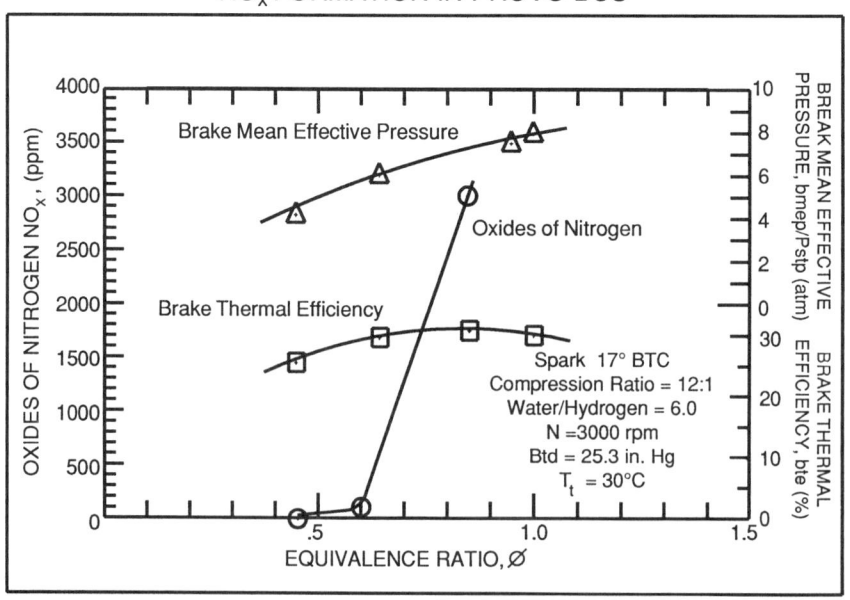

FIGURE 8-5: EFFECT OF EQUIVALENCE RATIO ON NO_x FORMATION IN PROVO BUS

hydride system mass of 1428 kilograms (3150 pounds). The bus metal hydride system is depicted in *Figure* 8-7. The actual amount of hydride alloy contained inside the stainless steel vessels was 1016 kilograms (2240 pounds), which could store 12.7 kilograms (28 pounds) of hydrogen, representing a 1.25 weight percent alloy and a 0.8 weight percent storage system. The metal hydride performance data is presented in *Figure* 8-8.

Cadillac Seville - 1977

As part of the "Hydrogen Homestead" project, this prototype was converted to hydrogen and operated as the family car. The engine conversion was accomplished with a modified Impco carburetor and water injection. The system was set up to allow the vehicle to be switched, while driving, back and forth from hydrogen operation to

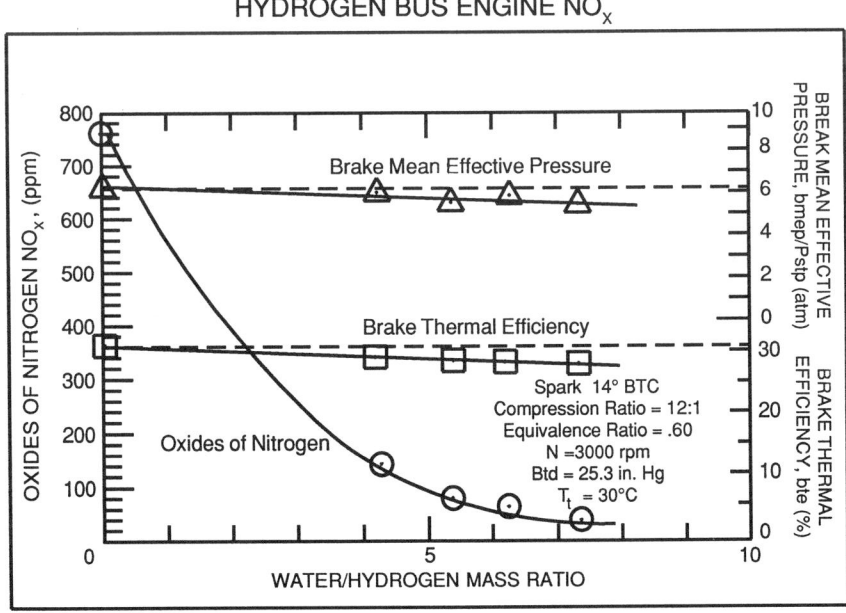

FIGURE 8-6: EFFECT OF WATER INJECTION ON HYDROGEN BUS ENGINE NO_x

gasoline operation. This was accomplished by leaving gasoline connected to the manifold fuel injectors which were electronically disabled during hydrogen operation, and by connecting a model 300A gas mixer above the engine's standard throttle plate to induct hydrogen into the system. Water injection was provided through water spray nozzles located in the intake manifold. Although the water injection was effective in controlling backfire in the engine, it was not properly distributed between the cylinders, resulting in nitric oxide concentrations in the exhaust of 25 to 100 parts per million.

The most important technical accomplishment of this prototype was the improvement of the metal hydride storage system. This prototype was much closer to a commercially manufacturable system than any of the earlier attempts. The system operated at a much lower temperature, deriving the heat needed to dissociate the hy-

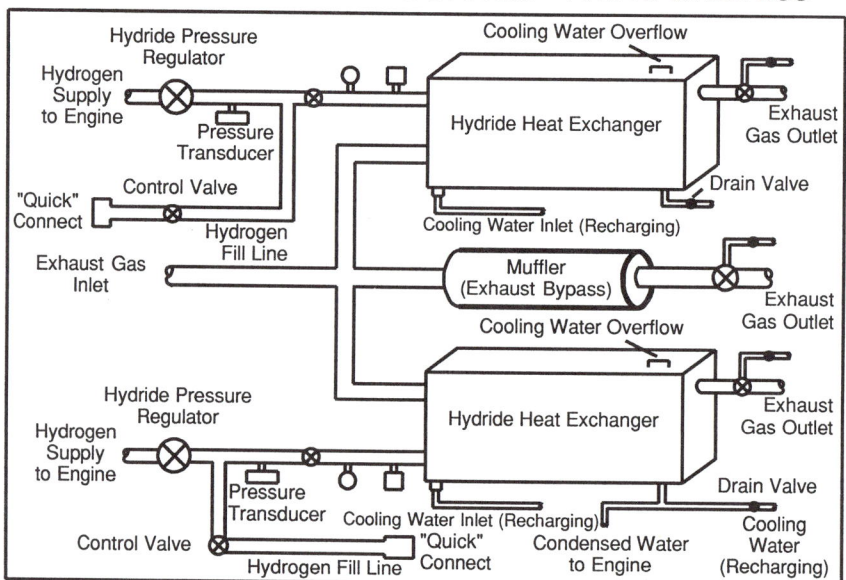

FIGURE 8-7: METAL HYDRIDE SYSTEM -- PROVO-OREM BUS

dride from engine cooling water. Since the system operated at a lower temperature, it was possible to fabricate the vessel out of carbon steel, resulting in a substantial cost reduction compared to the stainless steel predecessors. The heat from the cooling water was introduced into the steel vessel through an internal spiral coil made of copper tubing. See *Figure* 8-9.

From our earlier prototypes we had learned that with each recharge cycle, the Fe-Ti alloy decrepitates (continues to breakdown) into smaller and smaller particles. Our filter systems were set up to remove particles of 1 micron in diameter or larger. After many cycles, it was observed that the hydride alloy was becoming small enough to pass through the filters into the engine. To address this problem in the Seville prototype, we added a small percent of manganese to the Fe-Ti alloy. This procedure limited decrepitation to a particle size of 10 microns.

FIGURE 8-8: METAL HYDRIDE VESSEL PERFORMANCE
PROVO-OREM BUS

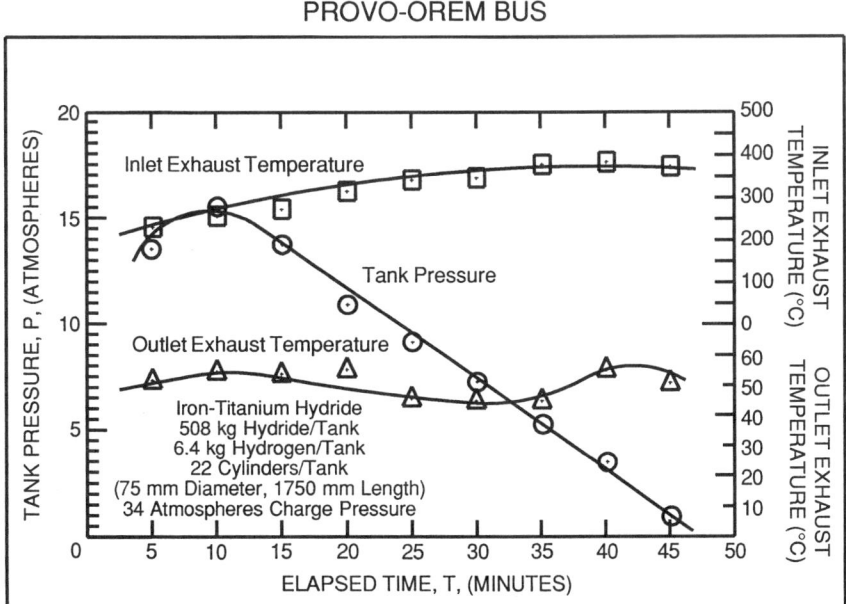

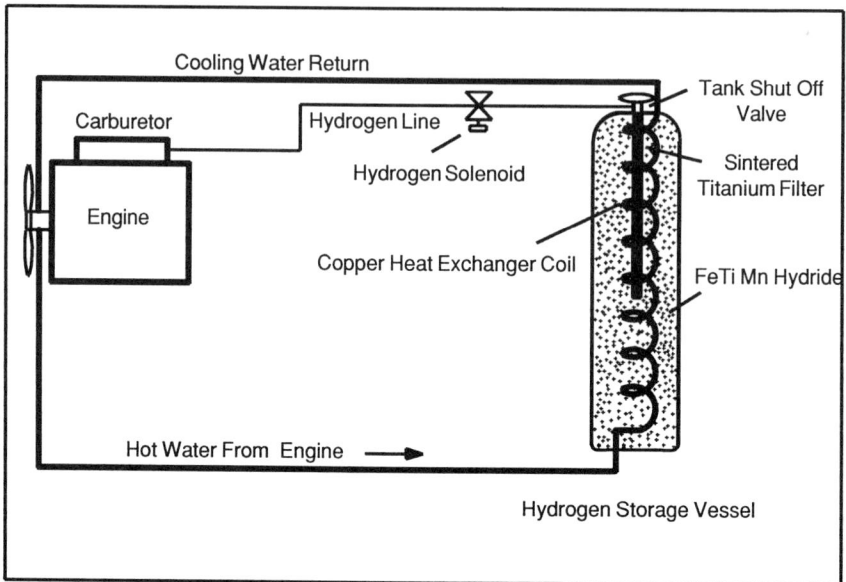

FIGURE 8-9: CADILLAC HYDRIDE SYSTEM

Jacobsen Lawn Tractor - 1977

The lawns at the "Hydrogen Homestead" were cared for with this converted mower. The hydrogen was stored in a metal hydride vessel utilizing the same technology as the Cadillac Seville. This vessel was mounted behind the driver over the rear wheels for added traction. The engine was manufactured by Kohler, one of the sponsors of the project. Since the engine was air cooled, the exhaust was passed through a heat exchanger which heated ethylene glycol solution which then passed through a coil internal to the hydride vessel to release the hydrogen.

Fuel Cell Lift Truck - 1977

Under contract from the military, a warehouse lift truck was modified for hydrogen operation. The lift truck was fitted with four phosphoric acid type fuel cells made by Engelhard. They were large and bulky, but the state of the art at that time. The hydrogen was stored in a metal hydride vessel which used the exhaust from the fuel cells to dissociate the hydride. The hydrogen then fueled the cells, with the resulting electricity powering the electric motors of the lift truck.

Riverside Bus - 1977

The Riverside, California hydrogen bus project was undertaken to demonstrate hydrogen's potential as a fuel for mass transit vehicles. Specifically, the bus was to be a second generation vehicle employing design improvements learned from operation of the first hydrogen bus in Provo, Utah. The most important improvement would be the development and installation of the direct cylinder injection (DCI) technology which would result in the elimination of backfire and would give the bus more power than when operated on hydrocarbon fuels. After completion of the development portion of the project, the bus would be operated to collect data documenting the pollution-free exhaust, the absence of normal engine wear, and the public's reaction to the new fuel.

This would be the first hydrogen vehicle to be operated and maintained by a public transit authority, a condition which turned out to be the single most important aspect of this experiment. Since the state transportation authority, Caltrans, was not willing to fund the project as part of their existing budget, the idea of the hydrogen bus project was presented before the California State Legislature. On January 27, 1975, Senate Bill Number 238 introduced by Mills and Ingalls was passed into law. Section 8 of the Bill appropriated $125,000 "for the purpose of converting a bus from conventional power to hydrogen power". In our proposal to the State of California, before passage of the Bill, we had requested $98,160 for development of the DCI engine system and $125,000 for conversion of the bus. Unfortunately, the $98,160 was not included in the appropriation. Without funding we could not develop a second generation vehicle having all of the advantages of a DCI system.

To make matters much worse, when the state turned the funds over to the City of Riverside for administration of the project, the City took out a giant chunk of the funding to pay their own expenses. Of the $125,000 which was appropriated to convert the vehicle to hydrogen, only $85,600 was left. After paying $24,000 for the purchase of a bus, there was only $61,600 to do the project, or just 28 percent of the required $223,160.

Against my better judgement, I went ahead with the project in spite of the lack of funding. I informed the city that the project, without adequate funding, would have to be scaled back to old technology, with the associated drawbacks (backfire and loss of power). They agreed to search for additional funding but none ever materialized. We worked hard on the project. A 19-passenger Argosy Bus was selected for conversion. The contract required that the bus have a hydrogen range of 97 kilometers (60 miles) of start-stop urban driving with a refueling time of 30 minutes or less. To meet these requirements it was determined that 907 kilograms (2000 pounds) of $Fe_{44}Ti_{51}Mn_5$ metal hydride alloy would be required. The heavy mass of the storage containers exceeded the capacity of the bus chassis.

Consequently, a special, heavy duty chassis was custom ordered. Still too heavy, based on the legal gross vehicle weight, we eliminated two passenger seats and made the metal hydride pressure vessels out of aluminum—the most important technical achievement of the project.

The hydride alloy was installed in 10 separate hydride vessels as shown in *Figure* 8-10. Each hydride vessel was 122 centimeters (48 inches) long, 20 centimeters (8 inches) in diameter, and made of aluminum. The mass of each vessel without the hydride was 22.6 kilograms (50 pounds). Around each tank was an external water jacket through which engine radiator water flowed to provide the heat of dissociation. See *Figure* 8-11. To accomplish the 30-minute recharge requirement of the contract, provision was made to connect a water line to the vehicle during recharging to cool the hydride tanks. The cooling system is shown in *Figure* 8-12. The tanks can be

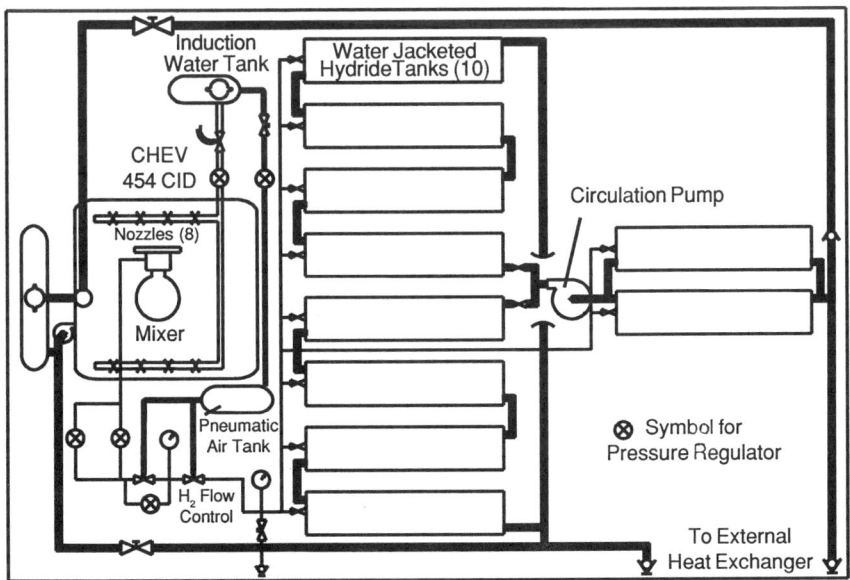

FIGURE 8-10: RIVERSIDE BUS HYDRIDE SYSTEM

FIGURE 8-11: HYDRIDE HEAT EXCHANGER DESIGN

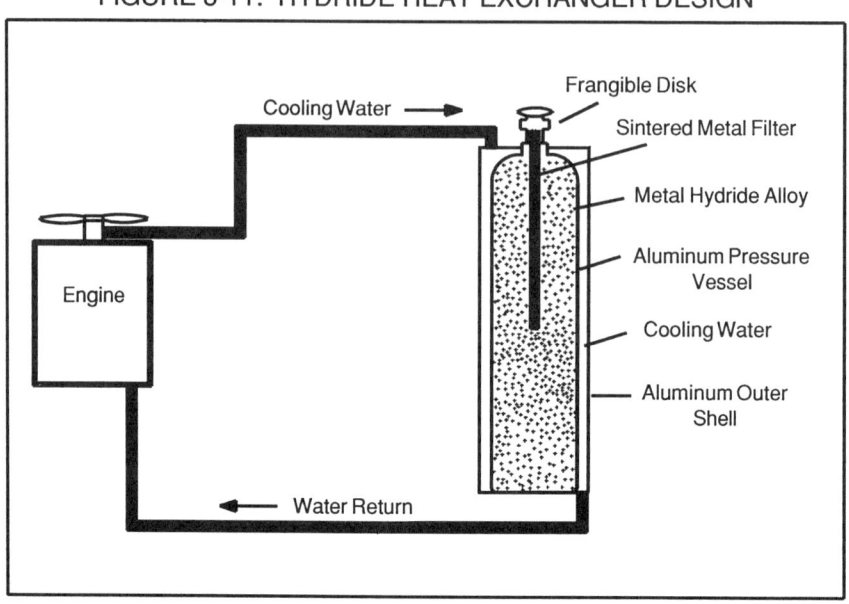

FIGURE 8-12: COOLING SYSTEM FOR FAST RECHARGE

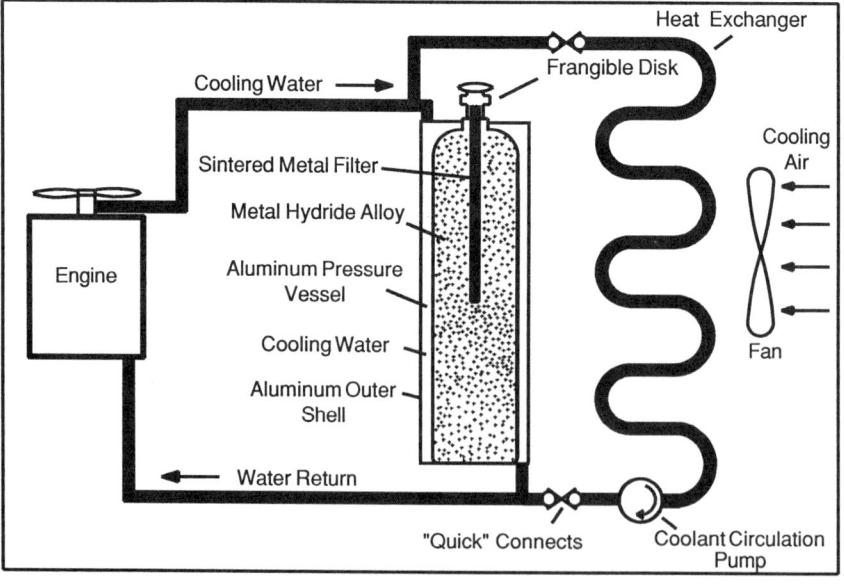

recharged very quickly if they are cooled, but require several hours to recharge if they are not cooled.

The real problems began when we delivered the bus to Riverside. Before the first day of scheduled operation of the bus, the engine suddenly began to malfunction. It lost most of its power and backfired repeatedly. I immediately flew to California from Utah with a team of technicians. After pulling the carburetor off of the engine we discovered a substantial quantity of a reddish "dirt"-like substance. A chemical analysis of the material revealed that it was silica, or sand. Since there was no silica used anywhere in the vehicle system, it was evident that the vehicle had been sabotaged. We took the bus back to Utah, fixed the engine, and returned it to Riverside. Then there were other problems. It seemed that someone was determined to make certain the project failed. Finally, to save the project, I sent a technician to California to protect and maintain the bus. Only then was the bus able to operate properly, but by now everyone involved in the project had become discouraged. Unfortunately, in spite of the circumstances, most of the blame for the problems was attributed to hydrogen. This was my first realization that not everyone shares my enthusiasm for a clean energy system. Powerful vested interests will not benefit from a shift to a new fuel. The opposition to the implementation of hydrogen energy has been fierce and often from high places. Someday, this story will be told.

Postal Jeep - 1977

Interested in the potential of a pollution-free fuel, especially in urban areas with a serious pollution problem, the U.S. Post Office contracted for the conversion of a Postal Jeep to hydrogen. As the operator of the largest fleet in the nation, the Postal Service was an ideal testing ground for the new technology. The project, which spanned several years, provided a substantial amount of technical experience resulting from the extended operation of a hydrogen vehicle.

The engine conversion method employed was an Impco carburetor with water injection. The hydrogen was stored in an iron-titanium-manganese hydride. In the first generation prototype, the hydride vessel consisted of aluminum cylinders with external water jackets similar to those in the Riverside bus. This design did not perform well in this application. The space occupied by the hydrides, which were mounted beneath the mail shelf, created inconvenience for the mail carrier and limited the amount of parcels that could be carried in the vehicle. The hydride system was replaced with a carbon steel vessel, made by welding together two tank ends. The resulting pressure vessel was the same dimension as the spare tire, and therefore could be mounted in a special cavity in the vehicle frame which had been made for that purpose. (The spare tire had already been relocated to the rear of the vehicle for convenience.) The heat exchanger for this vessel consisted of a network of copper tubing through which engine coolant was circulated, and which had been installed in the vessel before it was welded together.

In the final phase of the project, the vehicle was operated in Independence, Missouri, by a mail carrier. Careful data was collected to determine the efficiency of a hydrogen engine in a mail delivery application compared to a gasoline vehicle. To make the data meaningful, a control gasoline vehicle was operated in tandem with the hydrogen vehicle. Each day the miles traveled and the fuel

FIGURE 8-13: POSTAL VEHICLE FUEL EFFICIENCY COMPARISON

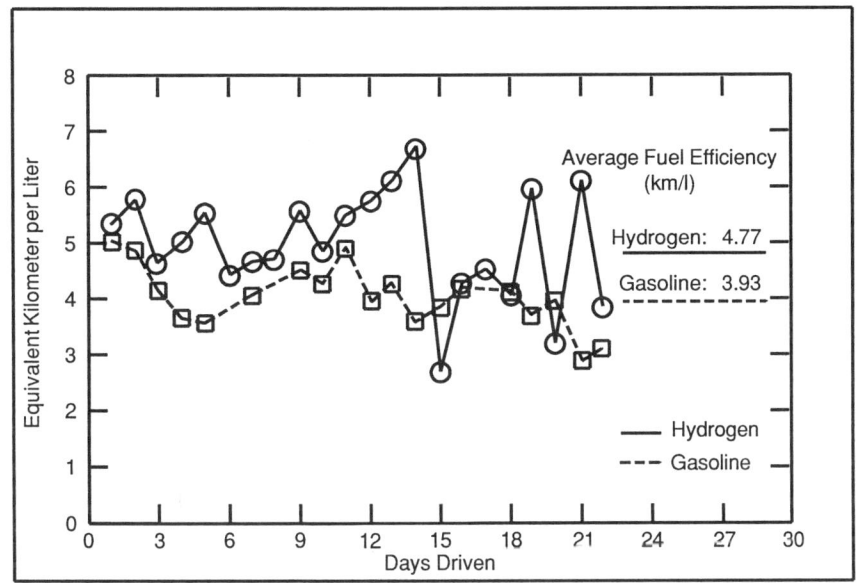

consumed was monitored for each vehicle. The results of these tests are presented as *Figure* 8-13.

The hydrogen-fueled vehicle proved more efficient than the gasoline counterpart, but the advantage was not sufficient to make up for the higher cost of hydrogen. We are now hopeful that the fuel cell technology will make up the difference, providing the Postal Service with a pollution-free alternative.

Los Alamos Airport Shuttle - 1981

A Buick Skylark was converted to hydrogen for the Los Alamos National Laboratory to shuttle guests to the laboratory from the airport. The engine was converted with a modified Impco carburetor and water injection. To store the hydrogen, a liquid hydrogen dewar was obtained from the DFVLR (German NASA) and installed in the vehicle.

Peugeot Sedan - 1982

The Peugeot prototype was converted for the factory in Paris, France. The standard gasoline vehicle was retrofitted with twin iron-titanium-manganese hydride vessels which were installed in the trunk. A map of the engine's performance on hydrogen was generated from data collected on an induction-type dynamometer. The engine conversion method utilized was direct cylinder injection with computer controlled, electronically actuated injectors.

LaserCel 1 - 1991 First Fuel Cell Automobile

The most recent prototype is the hydrogen fuel cell vehicle, LaserCel 1. Technical information concerning this prototype is included in Chapter 11.

Chapter 9

Hydrogen Production and Electrolysis

CHAPTER 9

HYDROGEN PRODUCTION AND ELECTROLYSIS

Hydrogen is the most abundant substance in our universe, and the third most common element on our planet. (One in every seven atoms is a hydrogen atom). Nevertheless, hydrogen in the free or unreacted state is very rare here on earth. Only a very small trace of hydrogen exists in air (less than 1 part per million). The bulk of hydrogen on the earth is in the reacted form as a constituent of water, hydrates, hydrocarbons, and organic (living) matter.

Actually, water is "burned hydrogen" or "hydrogen ash". Water is the by-product resulting from operating a vehicle on hydrogen. In more technical terms, then, water is a lower energy form of hydrogen. To turn water back into a fuel, energy must be pumped into the water causing it to dissociate, freeing the hydrogen. For this reason, we do not consider hydrogen to be a source of energy. It is, rather, an energy vector—a convenient form of energy that can be stored safely and then used efficiently without jeopardizing the environment.

Natural Gas Reformation

In the United States most hydrogen is produced by natural gas (methane) reformation. In the hydrocarbon reformation process, methane is reacted with steam over a nickel catalyst at a temperature of up to 1000° Centigrade (1800° Fahrenheit). The reaction for this process is:

$$CH_4 + H_2O \rightarrow CO + 3H_2$$

In the second stage of the reformation process, a reaction referred to as the "CO shift" occurs. In this reaction, carbon monoxide is reacted chemically with steam.

$$CO + H_2O \rightarrow CO_2 + H_2$$

This reaction takes place at 350° C (650° F) on the surface of commercially available catalysts (such as iron oxide). Oxygen from the steam combines with the carbon monoxide to form carbon dioxide, with the two hydrogen atoms in the water molecule being released as additional hydrogen. The resulting gas now consists of one molecule of carbon dioxide and four molecules of hydrogen. The overall chemical reaction for the reformation of natural gas is:

$$CH_4 + 2H_2O \rightarrow 4H_2 + CO_2$$

Fortunately, natural gas can be reformed into hydrogen without a substantial loss in energy. This is significant in view of the fact that hydrogen can be utilized in some energy applications much more efficiently than natural gas. It actually is beginning to make sense to convert the natural gas into hydrogen before utilization. More will be said about this topic in Chapter 11.

Coal Gasification

In addition to natural gas reformation, other hydrocarbons can also be transformed into hydrogen by the reformation process. More significantly, for the past 30 years commercial plants have been in operation to produce hydrogen by coal gasification. Normally, burning coal is dirty (such as in a coal fueled power plant). However, the same coal can be gasified into hydrogen with a technology that is efficient (approximately 80 percent) and which results in a substantial reduction in the pollution which would have been generated by direct coal combustion.

The chemistry of coal gasification is very similar to natural gas reformation. It consists of reacting steam with carbon in coal to form carbon monoxide and hydrogen. This low-BTU gas is then sent through a CO shift as was the case with methane reformation. The resulting products are hydrogen, carbon dioxide, and ash which is useful as a road base or cement additive. The basic reactions for coal gasification are as follows:

$$C + H_2O \rightarrow CO + H_2$$

$$CO + H_2 + H_2O \rightarrow 2H_2 + CO_2$$

In the coal gasification process, any sulfur present in the coal exits the gasifier as hydrogen sulfide. The hydrogen sulfide is removed by a Holmes-Stretford unit where the sulfide is absorbed and regenerated. The resulting sulfur is filtered out as a cake (39 percent by weight) which is sold as a valuable feedstock. This harmless and even useful by-product of coal gasification is in contrast to the toxic acid rain caused by sulfur dioxide resulting from the direct combustion of coal in an electric power generating plant. Coal gasification, then, is the cleanest way to use this vast hydrocarbon energy resource. Furthermore, coal gasified to make hydrogen will power three times as many vehicles as coal converted into other gaseous or liquid hydrocarbon fuels. (More detailed information on coal gasification is available in my book <u>Hydrogen From Coal - A Cost Estimation Guidebook</u>.)

The real need, of course, is to separate ourselves from hydrocarbon fuels. It is technologically possible to make the transition to hydrogen produced from renewable resources. Although such systems initially require a substantial capital investment, they will pay out over the long haul, and we should be moving in that direction as quickly as possible. Solar, wind, hydroelectric, and geothermal technologies are becoming sufficiently advanced for widespread commercialization. Hydrogen makes it possible to store these renewable energy resources, to transport them, and then to utilize them efficiently in actual energy applications.

Electrolysis

Utilizing a renewable energy source, water can be easily dissociated into its component parts, hydrogen and oxygen. The most simple process for dissociating water employs the use of electrical energy and is known as electrolysis. When two metal plates are placed in water in the presence of a catalyst and connected to a source of electricity, water molecules are pulled apart into hydrogen and oxygen. Hydrogen bubbles collect on the negative plate (cathode) while oxygen bubbles gather on the positive plate (anode). Since hydrogen and oxygen exist in water at a ratio of two to one, twice as many hydrogen bubbles form as oxygen bubbles. Equipment to commercially separate water into hydrogen and oxygen has been on the market for many years. This electrolysis equipment utilizes various schemes and technologies to increase the quantity of hydrogen produced per unit of energy consumed. The measure of hydrogen produced by an electrolyzer verses the electricity consumed is referred to as the electrolyzer's efficiency. If the amount of hydrogen produced by an electrolyzer were exactly equivalent to the electrical energy put into the unit, then the device would be said to be 100 percent efficient. In reality, commercial electrolysis equipment ranges in efficiency from 40 to 80 percent.

Each electron which is passed through water in an electrolysis device liberates one atom of hydrogen. Two electrons, then, produce one hydrogen molecule (H_2). Avogadro's number of electrons (6.02×10^{23}) produces one gram of hydrogen. Since each electron produces one hydrogen atom, the efficiency of an electrolysis device can be determined by measuring the electric voltage required to operate the cell. A cell operating at the theoretical voltage of 1.23 volts is 100 percent efficient. The amount of voltage above 1.23 required to operate the cell is wasted. The objective, then, is to make a cell that will operate as close to this voltage as possible.

Scientists have learned a lot about the electrolysis of water. For example, consider an experiment where two copper sheets (electrodes) are placed in a container of pure water and two volts of electric potential are applied across the electrodes. *Figure* 9-1 depicts such an experiment. Virtually no current is passed through the cell and no hydrogen is generated. Although the voltage is more than adequate, a couple of refinements need to be made to the cell.

First, it is necessary to add an electrolyte to the water. An electrolyte is a chemical which ionizes in the water and facilitates the conduction of electricity through the liquid. Very pure water has an extremely high electrical resistance (in excess of one megohm per centimeter). In a simple electrolysis cell, either hydrochloric acid (HCl) or potassium hydroxide (KOH) can be used as the electrolyte.

FIGURE 9-1: BASICS OF ELECTROLYSIS

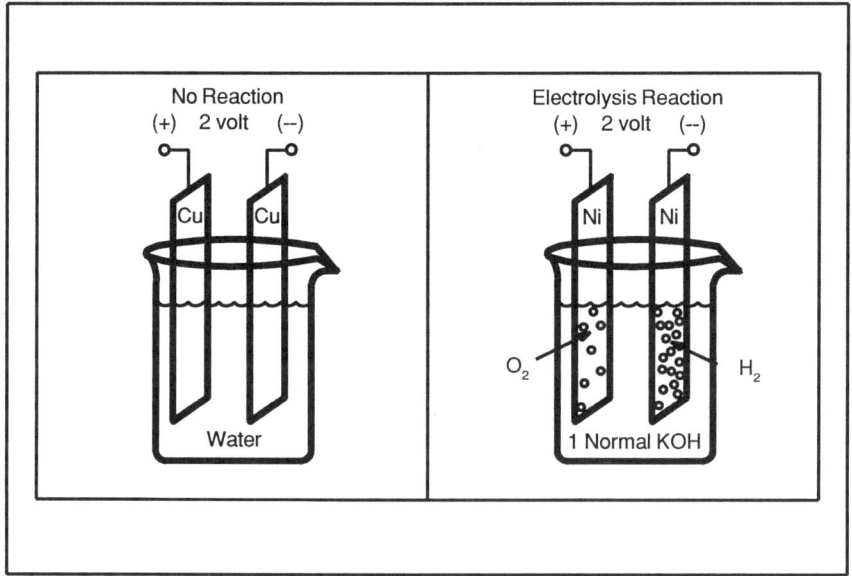

When even a small quantity of the electrolyte material is added to the water, the electric current begins to flow.

The next improvement to lower the voltage is a metallic catalyst. Instead of copper, which is not an ideal material for electrolysis, electrodes of nickel should be substituted. Now when connecting the circuit as shown in *Figure* 9-1, hydrogen bubbles begin to form at the negative electrode while oxygen bubbles form at the positive electrode.

Various techniques have been developed to commercially produce hydrogen electrolytically using this very chemistry. Typically, large sheets of nickel are immersed in water containing potassium hydroxide. Often asbestos cloth sheets are placed between the anodes and cathodes to prevent the hydrogen and oxygen bubbles from mixing together. The larger the surface area of the electrodes, the greater the current and consequently the greater the hydrogen and oxygen production. Typically, such a device will operate at a voltage of between 1.5 and 2 volts per cell. The higher the voltage the faster the hydrogen and oxygen evolution, but the greater the quantity of electricity wasted generating heat.

There are many drawbacks with this old technology, which has been in production for over 50 years. This type of equipment is very heavy and bulky. The hydrogen and oxygen are generated at near atmospheric pressure, so hydrogen must be compressed with an expensive external compressor. Most important, these units require extensive maintenance. As water is consumed in the process of electrolysis, additional water must continually be added to enable the process to continue. Over time, the impurities in the water begin to build up inside the cell. Any possibility of flushing these impurities out of the system is prevented by the KOH present in the water, which would then be lost. As a result, after many hours of operation these materials deposit on the inner surfaces of the electrolyzer and must be removed. To accomplish this, operation must be discontinued, and the electrolyte dumped. This process is time-consuming and expensive. It contributes substantially to the cost of hydrogen produced by this method.

Since 1974 I have been involved in the development of an electrolysis system in which the liquid electrolyte (KOH) is replaced by a solid electrolyte, a solid polymer electrolyte (SPE). A solid polymer electrolyte is a material in appearance very much like a thin sheet of transparent plastic. When the membrane is placed in water, however, it has very important chemical properties. The material is a perfluorosulfonic acid membrane. The electrodes coming in contact with the membrane encounter a highly acidic environment equiva-

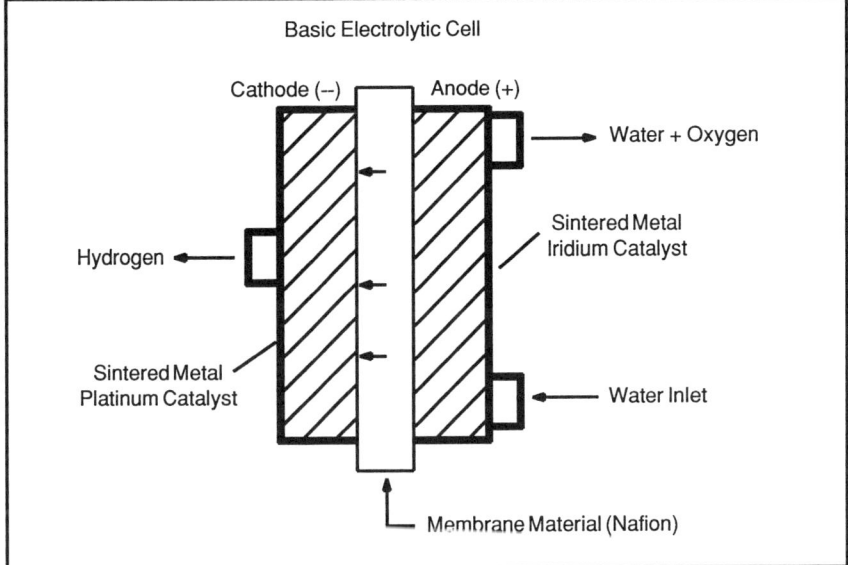

FIGURE 9-2: SPE ELECTROLYZER DESIGN

lent to a 20 weight percent sulfuric acid solution, in spite of the fact that only pure water is actually circulated in the system.

In operation, the membrane transports protons (a hydrogen atom without its electron) through the membrane in this manner: on the anode side of the cell, the electric current strips a hydrogen ion (proton) away from a water molecule and transports it through the membrane. On the cathode, the hydrogen ion receives an electron thereby evolving into hydrogen. A schematic diagram of an electrolyzer based on a solid polymer electrolyte is depicted as *Figure 9-2*. Water is also passed through the membrane with the hydrogen ion due to the electroosmotic effect.

Because of the highly acidic sulfonate environment of the membrane, acid-resistant noble metals are employed as the electrocatalysts.

Platinum is an ideal cathode material and contributes to highly efficient electrolysis. Most of the power loss, however, is at the anode. Here a phenomenon takes place known as "oxygen overvoltage". Oxygen overvoltage can be substantially reduced by coating the anode with a very thin layer of iridium. *Figure 9-3* compares the performance of an SPE electrolyzer with a platinum and an iridium anode.

SPE electrolyzers have many advantages over the older technology. First of all, they are very compact and lightweight. Second, it is easily possible to produce hydrogen in an SPE electrolytic cell at pressures as high as 40 atmospheres (600 psi). Since hydrogen is very difficult to compress because of its small molecular size, producing hydrogen already at pressure inside the electrolytic cell is a tremendous advantage. Third, since the electrolyte is solid membrane material

FIGURE 9-3: EFFECT OF IRIDIUM ON OXYGEN OVERVOLTAGE

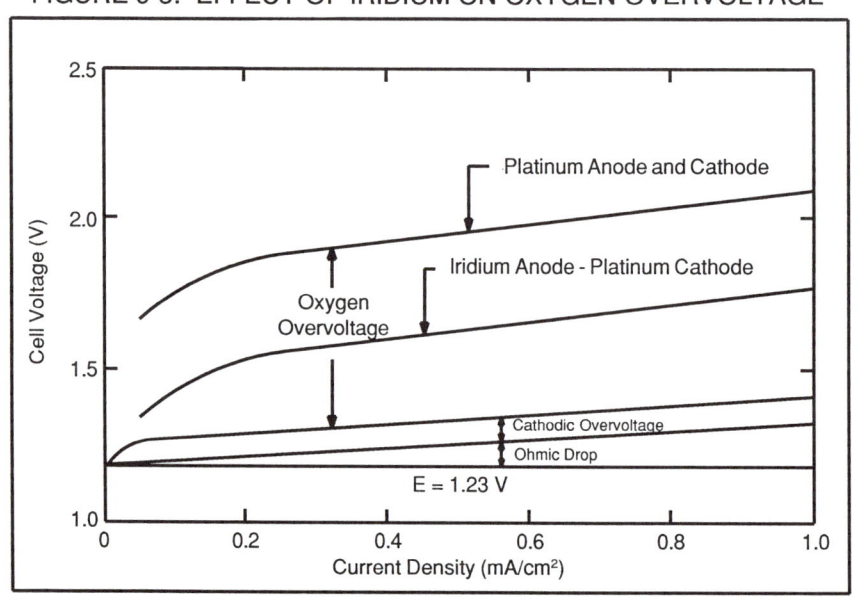

Hydrogen Production and Electrolysis 113

and not a liquid, it does not wash away. Consequently, in a solid polymer electrolyzer, a small amount of water is routinely drained from the cell. This procedure removes, on a continual basis, the concentration of impurities, greatly minimizing maintenance.

All that is necessary to build a cell with a solid polymer electrolyte is an anode and cathode which are placed on either side of the membrane, with a supply of water to the anode. *Figure* 9-4 is a schematic of an SPE electrolyzer system. The hydrogen exits the cell saturated with water vapor. If the gas is to be used to recharge a metal hydride storage vessel, a refrigeration device should be installed in the hydrogen line before the water trap to condense out the water. After the trap, the gas passes through a de-ox catalyst (available from United Catalysts) to remove oxygen by reacting it with hydrogen. Finally, the hydrogen is sent to a pressure swing

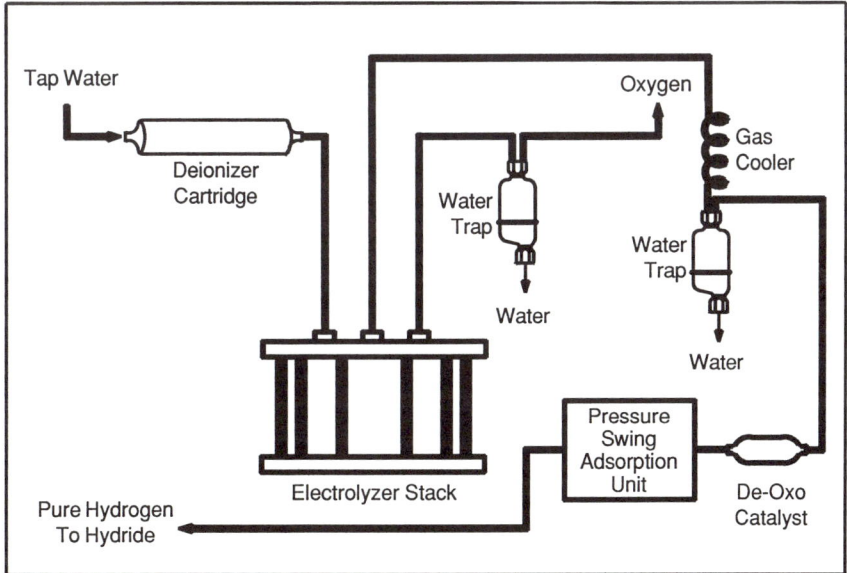

FIGURE 9-4: SPE ELECTROLYZER SCHEMATIC

adsorption (PSA) device where the remaining impurities are removed. (The PSA device utilizes a molecular sieve material such as that available through the Linde Division of Union Carbide to remove impurities from the gas.)

Because of their high efficiency (60-80 percent), SPE electrolyzers are ideal for converting renewable energy resources into hydrogen. The hydrogen can be generated utilizing electricity produced from a wide variety of energy sources. For example, photovoltaic solar collectors can be connected to an electrolyzer to generate hydrogen when the sun is shining. The hydrogen can be stored and then very efficiently utilized in a diversity of applications. Furthermore, hydrogen produced by solar-powered electrolysis can be combined indiscriminately with hydrogen produced from other energy sources. In this way, it will be possible to use hydrogen produced from a variety of energy sources as the energy vector of future society.

Biological Production of Hydrogen

In the future we can look forward to some very interesting and innovative new methods of producing hydrogen from renewable sources. For example, certain strains of blue-green algae have demonstrated the ability to separate water into hydrogen and oxygen, directly utilizing energy from the sun. Not only will such systems one day provide hydrogen to power our pollution-free communities, but the resulting blue green algae is an excellent food. How else can you "have your energy and eat it, too"?!

Chapter 10

The Hydrogen Homestead

CHAPTER 10

THE HYDROGEN HOMESTEAD

Hydrogen is an ideal fuel for application in a wide variety of devices. In 1975, this versatility was demonstrated in the Hydrogen Homestead project located in Provo, Utah.

The Hydrogen Homestead was a high-tech home in which the appliances were converted to operate on hydrogen. The hydrogen system employed in the Hydrogen Homestead is presented as *Figure* 10-1. Hydrogen for the project was produced by electrolyzing water. The electrolyzer utilized for the project was based on the SPE technology and was our first attempt to operate such a unit outside

FIGURE 10-1: HYDROGEN HOMESTEAD ENERGY SYSTEM

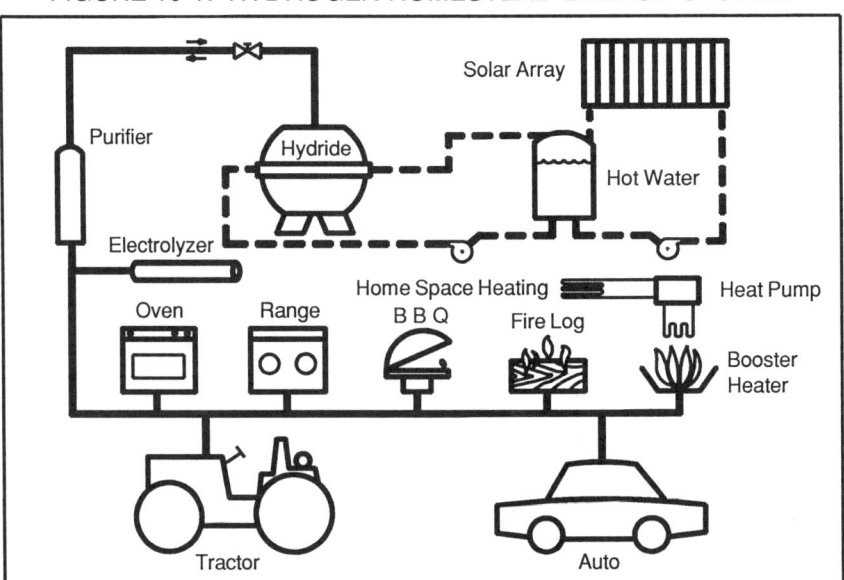

of the laboratory. The electrodes utilized in this device were based on a porous lead dioxide catalyst. Although the unit was very bulky and inefficient compared to later units, it did represent a major technological achievement.

As the hydrogen was produced, it was stored in an iron-titanium-manganese metal hydride alloy in the world's largest metal hydride vessel. This stainless steel vessel, which weighed 1791 kilograms (3950 pounds), is described in *Figure* 10-2. Because of its similarity in appearance to the famous Star Wars robot, this vessel was lovingly dubbed "R2D2". The vessel was instrumented with temperature probes, micro-expansion measurement transducers, and an internal vessel viewing window—all for the purpose of evaluating the feasibility of metal hydrides for bulk hydrogen storage. The stainless outer vessel was encapsulated in a 4-inch layer of foam to insulate it from the external environment. Heating and cooling of

FIGURE 10-2: HOMESTEAD METAL HYDRIDE VESSEL

Height	123.2 cm (48.5 in)
Diameter	97.3 cm (38.3 in)
Wall Thickness	2.38 cm (.937 in)
Vessel Material	Mild Steel: 0.15 w/o C, 1.1 w/o Mn
	0.09 w/o Cr, 0.005 w/o N
Internal Volume	.5969 m^3 (21.08 ft^3)
Service Pressure	3447 kPa (500 psig)
Test Pressure	6895 kPa (1000 psig)
Hydride Composition	51% Ti, 44% Fe, 5% Mn
Hydride Mass	1791 kg (3950 lb)
Service Temperature	55°C (131°F)
Pressure Excursion	7-3447 kPa (1-500 psig)
Usable Weight Percent	1.72%
Stored Hydrogen	30.81 kg (67.92 lb)
Stored Energy (Higher Heating Value)	4.37 GJ (4.14 million Btu)

the vessel was accomplished with an internal heat exchanger. The vessel stored 31 kilograms (68 pounds) of hydrogen, a quantity adequate to maintain operation of the Hydrogen Homestead for about 10 days.

The natural gas range was modified for operation on hydrogen. To accomplish this it was necessary to determine whether or not the size of the orifices needed to be changed. Hydrogen contains approximately one third the energy per unit volume of natural gas. That means for the appliance to operate on hydrogen at about the same energy level it did on natural gas, it was necessary for a volume of gas three times as great to flow through the burner orifice. Fortunately, hydrogen has a very low viscosity, caused by its extremely small molecular size. For this reason hydrogen flows through an orifice approximately three times as fast as does natural gas, meaning that natural gas orifice sizes are suitable for hydrogen conversions without any modification.

Hydrogen does have a very high laminar flame speed and virtually no suitable burner quench diameter. Let me explain. In a conventional natural gas burner, the velocity of the gas passing through the orifice draws air into the burner. The quantity of air is usually adjustable with a metal plate which can be turned to change the size of the inlet air holes. Air which enters the burner in this way is called primary air. The natural gas and air mix together as they travel up the length of the burner to the burner head. The burner head is fabricated with small openings through which the methane/air mixture passes. These openings are designed to be of such a size that the flame cannot pass through them back into the internal portion of the burner. The size of hole which is small enough to prevent the flame from passing back into the internal portion of the burner is referred to as the "quench diameter". In the case of hydrogen it is almost impossible to make the hole small enough to quench a hydrogen flame. In the first place, the quench diameter for hydrogen is much smaller than it is for natural gas. In the second place, when the metal out of which the burner is made begins to get hot, it acts like a catalyst to ignite the hydrogen on the inside of the burner. Consequently, the flame gets through even if the diameter of the holes is small.

This problem could be prevented by making the burner head out of very thick material with very small holes and by somehow conducting the heat away from those holes so that the inner surfaces do not reach catalytic ignition temperature.

Fortunately, however, none of these complicated adjustments are necessary. A much simpler solution is available. Hydrogen has such a high flame speed that it is better not to pre-mix the hydrogen and air as is done in the case of natural gas. The first step in converting a burner to operate on hydrogen then, is to eliminate the primary air intake. In some cases this can be done by adjusting the inlet air openings to zero. On other devices, the adjustment cannot be made, and it is then necessary to block off the inlet primary air passages by some other method. However this is achieved, it is important that primary air be eliminated.

Secondary air in a natural gas appliance is that air which mixes with the fuel and supports combustion after the fuel has passed through the burner head. (See *Figure* 10-3.) In the case of a hydrogen burner, secondary air is more than adequate for combustion. If natural gas is burned without primary air, a yellow flame results. The yellow flame is an indication that the natural gas is not being completely combusted. Such a flame will cause carbon buildup on cooking pots and puts dangerous quantities of carbon monoxide into the air. In

FIGURE 10-3: NATURAL GAS BURNER DESIGN MODIFICATIONS

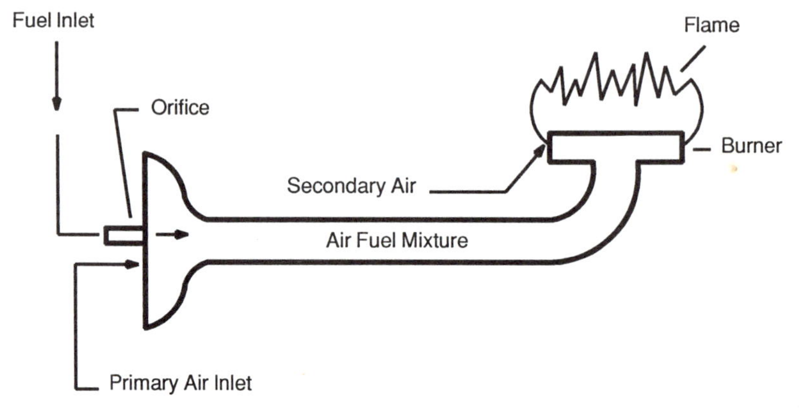

Diagram showing: Fuel Inlet, Orifice, Primary Air Inlet, Secondary Air, Air Fuel Mixture, Burner, Flame

natural gas appliances, the primary air inlet is adjusted until a blue flame is achieved, thereby indicating that efficient, complete combustion is taking place. Such flames are virtually free of carbon monoxide production since all of the carbon monoxide is oxidized into carbon dioxide (CO_2).

In a hydrogen burner there is no carbon, and no problem with incomplete combustion. In fact, the problem with hydrogen is the other extreme. Even without primary air, the hydrogen burns too quickly. This results in regions of high temperature in which nitrogen and oxygen—the constituents of air—combine to form nitric oxide. In a conventional range burner, we measured a nitric oxide concentration of 40 parts per million when the burner was operated on natural gas. When we converted the burner to operate on hydrogen by eliminating the primary air, the nitric oxide concentration jumped to 250 parts per million. Either concentration is too

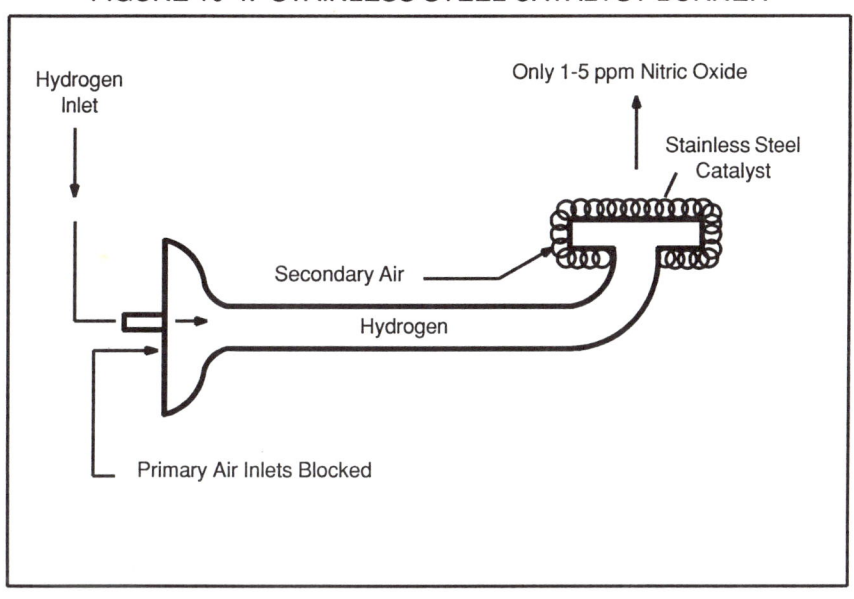

FIGURE 10-4: STAINLESS STEEL CATALYST BURNER

high when you consider that these stove-top burners are operating in the enclosed environment of your home.

Fortunately, a simple solution is available to eliminate this problem for hydrogen. At high temperature, stainless steel is an excellent catalyst for hydrogen combustion. Utilizing this property, a simple technique can be successfully employed to control nitric oxide formation. The stainless steel catalyst can be obtained by purchasing a stainless steel scouring pad from the grocery store. This catalyst (scouring pad) is carefully fitted around the outside of the burner head. See *Figure* 10-4. The catalyst performs three functions: First, it inhibits the mixing of hydrogen and air, thereby causing a zone immediately around the burner head where the concentration of hydrogen is very high and the concentration of air is very low. Second, it catalytically supports the reaction between hydrogen and air at concentrations which are not flammable. In other words, the hydrogen and oxygen are combined on the surface of the catalyst at a slower rate, thereby eliminating the reaction zone where the high temperature is normally generated. Since there is no high temperature zone, nitric oxide formation is virtually eliminated. Third, and an additional benefit of this approach, it produces a visible flame. Normally when hydrogen burns, the flame is invisible. This could be dangerous to the housewife who might think that the appliance is turned off when, in fact, it is in operation. The stainless steel catalyst glows a bright orange, making it apparent that the flame is burning while, at the same time, giving some indication of how high the burner is set.

Catalytic burners were utilized in virtually all of the natural gas-to-hydrogen conversions employed in the Hydrogen Homestead. Careful measurements indicate that only one to five parts per million of nitric oxide are generated by such a burner.

FIGURE 10-5: HYDROGEN WATER HEATER CONVERSION

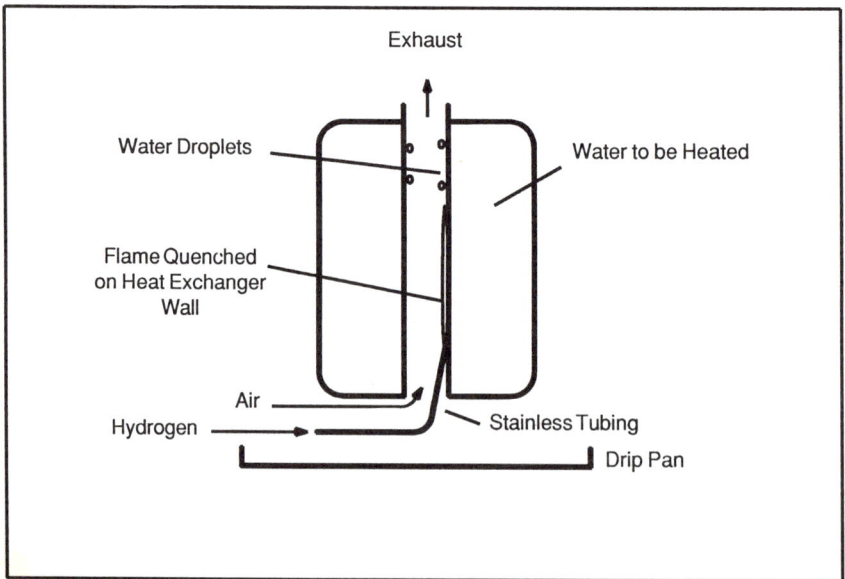

One very significant advantage of using hydrogen instead of conventional fuels, which was demonstrated by the Hydrogen Homestead experiments, was the increase in utilization efficiency when appliances were converted to hydrogen. Consider the hydrogen water heater. When this appliance was converted, the stainless steel catalyst was not necessary because it was possible to control the mixing of hydrogen and air in such a way that the actual combustion took place on the surface of the water heater's internal heat exchanger. This was done simply by designing a burner which consisted of a piece of stainless steel tubing which sprayed the hydrogen against the metal wall of the heat exchanger. (See *Figure 10-5.*) Since the metal is a very good conductor of heat and is constantly being cooled by the water on the other side of the wall, the flame was very effectively quenched with virtually no nitric oxide being formed. (Nitric oxide levels of 3 parts per million were measured from this appliance.) In the hydrogen water heater, it was possible to actually cool the steam in the exhaust gases until it condensed into droplets of water which would drain down the sides of the internal heat exchanger and drip into a drain pan underneath. It takes a lot of energy to vaporize water into steam—but when steam is condensed back into water, that same amount of energy is recovered. In this case, the energy liberated by condensing the steam went into heating the water inside the water heater, thereby greatly increasing the efficiency of the appliance. With the hydrogen-fueled water heater we were able to take advantage of the higher heating value of the fuel.

This advantage is also possible in the conversion of the furnace. During the cold months of the year, the humidity inside of a house becomes uncomfortably low. Expensive heating systems have a power humidifier to put additional moisture into the air, thereby increasing the comfort level. Since the only by-product of burning hydrogen (with the stainless steel catalyst) is pure water vapor, it was possible to modify the furnace so that all of the exhaust gases went directly into the house. This eliminated the waste of heat going up the chimney and also took advantage of the higher heating value of the fuel. All of the steam was captured within the living environment of the home, increasing the humidity and lowering the fuel bill. (Isn't hydrogen great?!)

The Hydrogen Homestead was also equipped with a hydrogen fireplace log, a hydrogen barbecue grill, a hydrogen lawn tractor, and a hydrogen-fueled Cadillac Seville. A lot of the problems associated with implementing hydrogen energy in a residential setting were encountered and overcome during the Hydrogen Homestead project. Most importantly, it was determined that hydrogen was compatible with existing home appliances. (The only real surprise we had in the world's first hydrogen home were the "hydrogen enthusiasts" we found peering expectantly in through the windows, looking for a "better world". We solved this "problem" by enlisting their help to make one!)

Chapter 11

The Hydrogen Fuel Cell Efficiency is the Key

CHAPTER 11

THE HYDROGEN FUEL CELL EFFICIENCY IS THE KEY

As we begin to move closer to a world in which vehicles are fueled by hydrogen, some real technological opportunities become available to us as a result of the unique chemical properties of hydrogen. One of the most interesting of these new technologies is the hydrogen fuel cell.

Internal combustion engines, although capable of being modified to operate on clean-burning hydrogen, waste more energy than they actually utilize for propulsion. It is well known that most of the energy generated by combusting fuel inside an internal combustion engine is rejected to the environment through the engine radiator or is wasted out the tailpipe as hot exhaust gases. Only 23 percent is actually used to power the vehicle. Although hydrogen conversions cause significant increases in the efficiency of internal combustion engines, there is still a much better and more efficient way to utilize the hydrogen—in a hydrogen fuel cell.

A hydrogen fuel cell is an electrolytic cell in which hydrogen and oxygen combine to form water. The resulting energy is liberated as electricity. In other words, a fuel cell is an electrical generator which is powered by hydrogen. Unless a blower is used to supply air to the fuel cell, it has no moving parts, it makes no noise, has absolutely no air pollution, and utilizes the hydrogen fuel very efficiently.

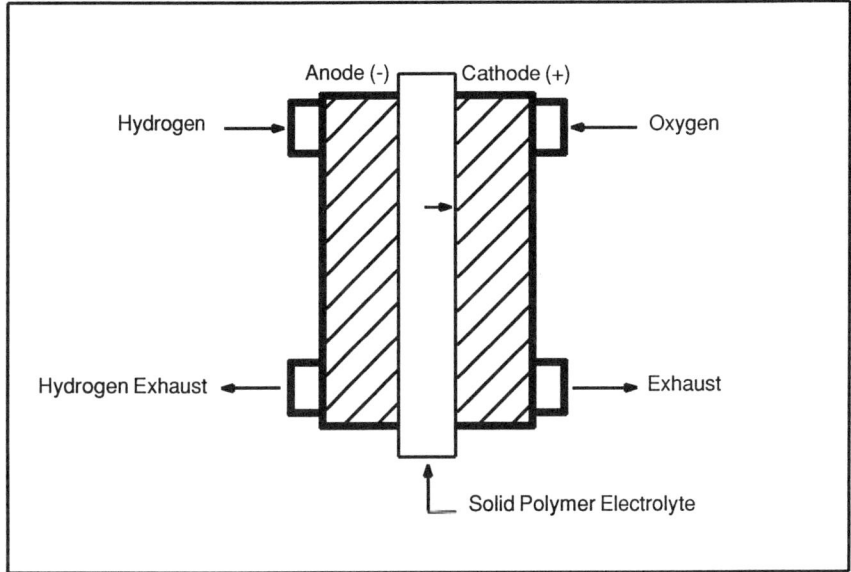

FIGURE 11-1: SPE FUEL CELL

Figure 11 1 is a diagram of an SPE fuel cell. In the fuel cell, a solid polymer membrane is located between two electrodes. On the side of the negative electrode (the anode), hydrogen gas is supplied to the fuel cell. On the opposing positive side (the cathode), oxygen is supplied. The anode reaction is:

Anode Reaction $H_2 \rightarrow 2H^+ + 2e^-$

Hydrogen dissociates in the presence of the catalyst, forming hydrogen ions (protons) and giving up electrons to the anode. The hydrogen ions are transported across the membrane to the cathode. At the cathode, hydrogen ions react with oxygen to form H_2O.

Cathode Reaction $2H^+ + 2e^- + 1/2\, O_2 \rightarrow H_2O$

The electrons, which are liberated at the anode, travel by electrical cable through the external load (such as an electric motor) to the cathode where they become part of the water-forming reaction. If the external circuit is "open", not allowing the electric current to flow, then the reaction is stopped. No fuel is consumed and no power is generated. The electrolytic reaction, then, is controlled by the load connected to the cell.

The overall fuel cell reaction is:

$$H_2 + 1/2\ O_2 \rightarrow H_2O$$

By this simple chemistry, each cell generates a maximum potential of just over one volt. The cells are stacked in series to achieve higher voltages as required.

Figure 11-2 shows a typical fuel cell output versus current density. As can be seen from this figure, the output voltage of a fuel cell

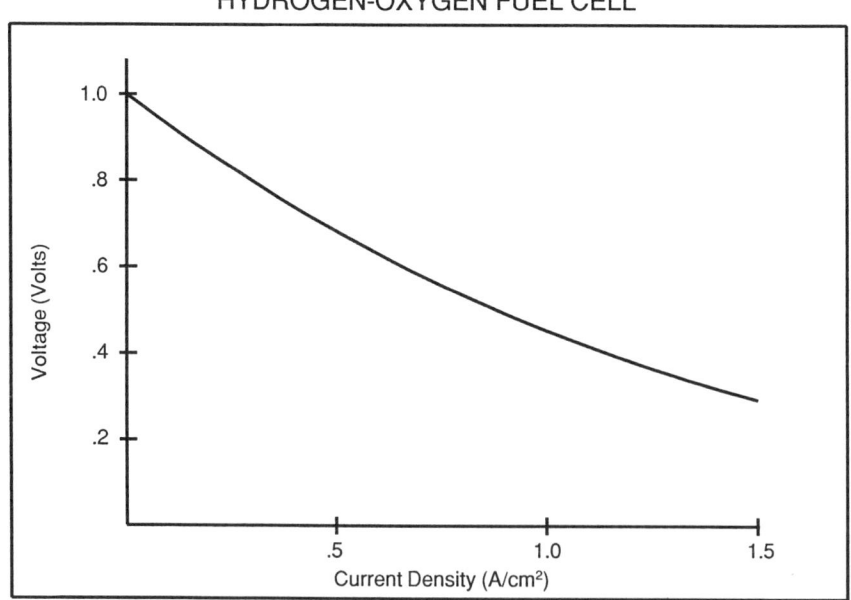

FIGURE 11-2: SPE FUEL CELL POWER CURVE
HYDROGEN-OXYGEN FUEL CELL

decreases as the electrical load is increased. The theoretical polarization voltage of 1.23 volts per cell (at no load) is not realized in real life due to various losses. Typically, solid polymer electrolyte fuel cells operate at .75 volts per cell under peak load conditions or at about a 60 percent efficiency.

The efficiency of a fuel cell is a function of such variables as catalyst material, operating temperature, reactant pressure, and current density. Typical operating efficiencies of around 60 percent are normal, but at low current densities, efficiencies as high as 75 percent are achievable.

A fuel cell, then, is a device which converts hydrogen and oxygen into water while efficiently producing electrical power. Some fuel cells have the capability of being operated in reverse as an electrolyzer. The dual-role capability of these devices makes them a contender for off-peak electric power storage applications.

In the case of the fuel cell car, it is possible to operate the fuel cell in reverse at night by connecting a source of electricity and water to the unit. In this electrolysis mode, hydrogen is generated which is used to recharge the hydrogen storage vessel. The next day, the electric power and water are disconnected, and the unit switches over to the fuel cell mode, generating electricity to power an electric motor which, in turn, provides the source of propulsion for the vehicle.

This capability is extremely important as we begin to implement hydrogen-fueled vehicles. The fuel cell utilizes hydrogen up to three times as efficiently as does an internal combustion engine. This increase in utilization efficiency has the effect of extending the vehicle range to 700 kilometers (450 miles) without increasing the weight of the metal hydride storage system. Even more important, the higher efficiency of the fuel cell lowers the cost per mile for hydrogen to one-third of the cost of fuel for an internal combustion engine. That makes the fuel cost less per mile for hydrogen than for gasoline. The fuel cell makes hydrogen-fueled vehicles practical and economical for the first time. Even hydrogen produced utilizing expensive solar collectors is now competitive with the present price for imported hydrocarbon fuels. The higher utilization efficiency of hydrogen fuel cell cars is the long sought KEY that makes the hydrogen energy dream come of age. At last we can compete. At last we can get started. No longer must we wait for an infrastructure. Hydrogen fuel cell cars can be recharged overnight by the existing electric grid.

The efficiency advantage of the hydrogen fuel cell system must be taken into account in considering today's popular movement towards retrofitting internal combustion engines to operate on methane as an answer to the pollution/energy crisis. While there are several arguments in favor of converting automobiles to operate on LNG or CNG (liquified or compressed natural gas)—for example, these cars produce considerably less pollution than their gasoline or diesel counterparts, and, at least for the short term, we have an adequate supply of this inexpensive fuel—we must take into consideration the fact that fuels (even hydrogen) burned in an internal combustion engine are largely wasted. Typically, 75 percent of the energy derived from burning a fuel in an internal combustion engine is wasted as hot gases escaping through the exhaust system or as heat lost to the environment through the engine radiator. Only one fourth of the energy is actually utilized to turn the wheels.

It becomes apparent, then, that there is a major advantage to converting the natural gas into hydrogen and then using the hydrogen in fuel cells where 60 percent of the energy from the hydrogen is utilized to power the electric motor and turn the wheels. If the

FIGURE 11-3: COMPARISON OF VEHICLE RANGE AND CO_2 EMISSIONS FOR METHANE- AND HYDROGEN - FUELED VEHICLES

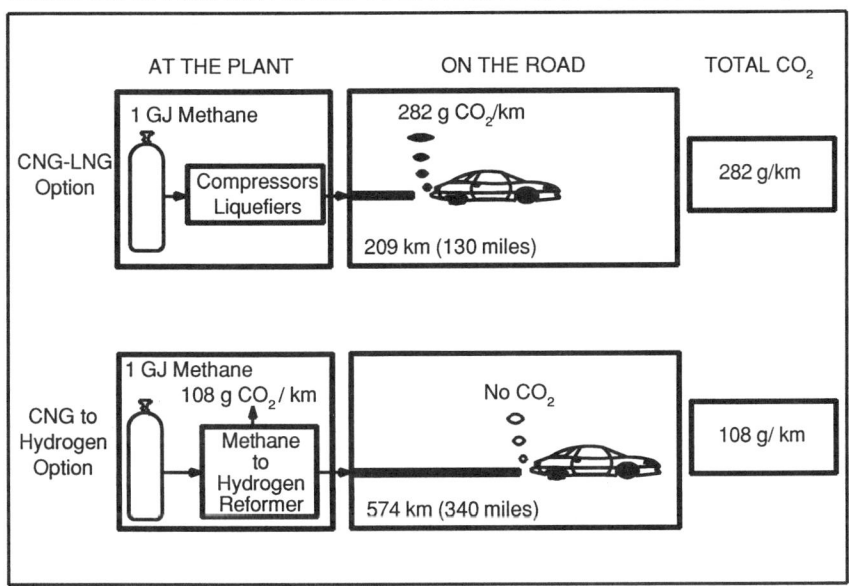

natural gas necessary to power a fleet of 1000 cars were reformed into hydrogen and then utilized in hydrogen fuel cell cars, the same amount of gas would power a fleet of 3000 vehicles. This means that only one third as much carbon dioxide would be released into the atmosphere. What a tremendous first step towards combating the "greenhouse" effect. *Figure* 11-3 compares the use of one gigajoule of natural gas directly as a fuel and then converted to hydrogen to power a fuel cell vehicle. As a fuel, the natural gas will provide 209 kilometers (130 miles) of driving and will release 282 grams of carbon dioxide per kilometer. By reforming the same amount of natural gas into hydrogen for a fuel cell vehicle, the range is extended to 547 kilometers (340 miles) and the carbon dioxide is reduced to just 108 grams per kilometer.

Hydrogen fuel cells have been around for many years. They were used in the Apollo program which placed man on the moon.

Presently fuel cells are used as the source of electrical power on the Space Shuttle. These fuel cells used in space have been very expensive, very bulky, and unreliable. My first fuel cell vehicle project was a phosphoric acid fuel cell conversion for a MERADCOM lift truck in the late seventies. The old type phosphoric acid fuel cells were not applicable to an automobile for many reasons. They were too bulky to fit under the hood, too unstable to be reliable, too slow at start-up, and, most important, way too expensive.

The answer was the development of a fuel cell based on solid polymer electrolyte technology. Our first generation technology SPE fuel cell improved considerably upon the phosphoric acid technology, but was still far too bulky and expensive for an automotive application. After selling my sixty-percent interest in Billings (Energy) Corporation, I decided to focus my full attention on the development of such a device. Funding the research with my own money, I assembled a research team at the American Academy of

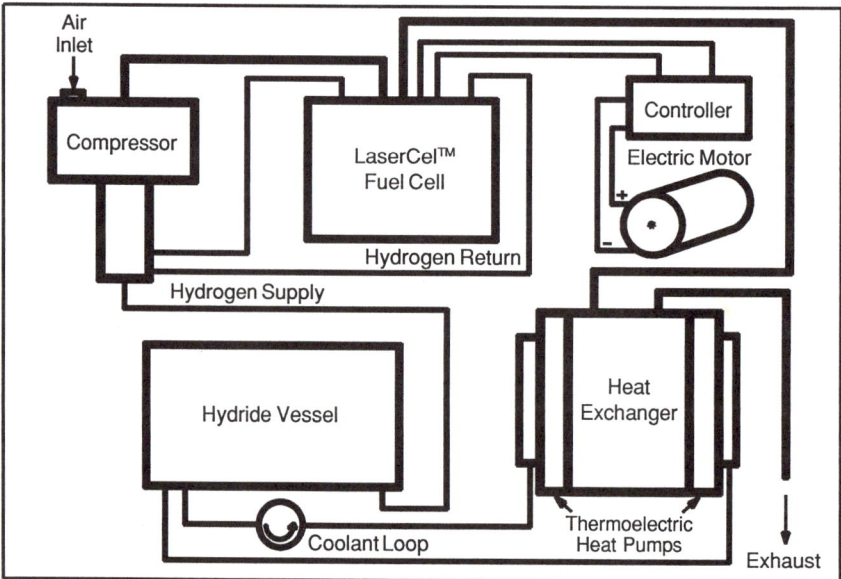

FIGURE 11-4: FUEL CELL VEHICLE BLOCK DIAGRAM

Science. The breakthrough came when we learned how to fabricate an advanced, double-sided anode/cathode design with the use of a high-powered laser. In our new fuel cell design, a solid polymer electrolyte replaces the older phosphoric acid technology. These new devices are capable of much higher current densities and, therefore, are much more compact and affordable than the earlier units.

The first automobile to employ a hydrogen fuel cell as its source of power was put on the road in 1991. We call this prototype "LaserCel 1". A schematic diagram of the LaserCel 1 prototype is depicted in *Figure* 11-4. Hydrogen is stored in the metal hydride storage vessel for safety. Upon demand, the fuel is released by the hydride and travels to the fuel cell in gaseous form. Inside the fuel cell, the hydrogen is reacted with air to generate electricity, with pure water vapor as the only by-product. The electricity is then supplied to an electric motor which is coupled to the front wheels of the vehicle. To provide electrical energy for fuel cell start-up and for periods of rapid acceleration, an accelerator battery is included as part of the system. Without an accelerator battery, it would be necessary to build the fuel cell much larger than is otherwise necessary. It is much more efficient to provide an accelerator battery to the system than it is to build a fuel cell capable of producing the peak power requirement of the vehicle. The accelerator battery supplies power to the motor during times of fast acceleration, but during the rest of the driving cycle, excess power generated by the fuel cell is used to recharge the accelerator battery.

At the time of this writing, the concept of fuel cell powered automobiles is very new. In fact, the critics still do not believe the fuel cell vehicle is real. Eventually, however, I believe that this prototype will become recognized as the long-awaited breakthrough technology that makes hydrogen vehicles really feasible.

The fuel cell is also the key to other hydrogen energy applications, such as the storage of off-peak electric power. In the large, densely populated urban areas, it is very difficult for electric utilities to supply adequate electric power to all of their customers during times of peak demand. Not only is it difficult to generate the excessive amounts of energy consumed at such times, but it is also difficult to distribute the power through the existing electric power network. During other parts of the day (i.e. while everyone is asleep), the electric generating capacity of the electric utilities is not fully utilized. One future application of fuel cells will be to use this off-peak electric power generating capacity to power fuel cells operating in the electrolysis mode, storing up quantities of hydrogen energy throughout the electric utility grid. During times of peak demand, these fuel cells will then begin to consume the generated hydrogen to produce electric power without pollution, without noise, and right at those places within the electrical grid where the power is really needed.

Taking the idea of the fuel cell a step further, I envision hydrogen-fueled homes of the future where natural gas or hydrogen will be distributed to the homes via underground pipelines. If natural gas, it would then be reformed into hydrogen and used to power a fuel cell, which would be the source of electrical energy—either for a single residence or for the homes of an entire neighborhood. Looking even further ahead to a time when only renewable sources of energy will be utilized, solar panels and wind collectors will generate electricity, which is converted into hydrogen by the fuel cell operating in the electrolysis mode, and stored. Then, at night or when the wind is not blowing and when the homeowner wishes to use electricity, the fuel cell will convert the stored hydrogen back into electrical energy. By this method, it is technologically feasible to begin thinking about energy independence at the individual residence level.

Chapter 12

Our Hydrogen Energy Future
How We Begin

CHAPTER 12

OUR HYDROGEN ENERGY FUTURE
HOW WE BEGIN

Although this book has described some very interesting technologies and various prototypes built to demonstrate the feasibility of utilizing those technologies, still fossil fuels reign king with continued oil spills, air pollution, global warming, and even armed conflict resulting. Until hydrogen energy begins to emerge in the commercial marketplace, displacing the conventional fuels and remedying our energy problems, all of my work with hydrogen energy and the efforts of thousands of colleagues have been to no avail. We must find ways to get this technology off the ground.

Having wrestled with these problems now for twenty-five years, I have formed some opinions on what steps must be taken to get started with the long-awaited implementation phase of this important project.

Step 1. First we must develop scientific solutions to the technical problems which stand as impediments to the commercial launching of hydrogen energy. Some of these problems concern the practical and safe utilization of the fuel. For example, we found that until we could safely store hydrogen on board a vehicle, it was very difficult to get anyone to seriously consider manufacturing such a product. Other problems still needing to be solved are associated with marketing questions such as convenience, performance, and cost. Scientists must make the new technology better than the existing options before full-scale implementation can get under way. Exactly how good our solutions need to be depends on the amount of support available for the new technology and the urgency forced upon us by problems associated with the continued use of fossil fuels.

For some time scientists have explained the problems—long-term—associated with our continued use of fossil fuels. And although the public, by and large, believes these cries of gloom and doom, the warnings have proven to be insufficient incentives to get anyone to really do anything. With metal hydrides and the fuel cell technology becoming available, I believe we have the scientific, technical solutions which are now completing this first step towards commercial implementation.

Step 2. The next step is to build prototypes of the new technologies and to perform testing to optimize the technology and to obtain the necessary technical documentation which will be so very important during the next step towards implementation.

Fortunately, some of the necessary prototype testing has been ongoing over the past several years. Much remains, however, to be done. This is a very expensive part of the overall undertaking, and reliable sources of substantial amounts of funding must be found.

Understandably, the U.S. Federal Government has not been a good source of funding for these types of projects. The "vested interests" which may not directly benefit in the beginning phases of implementing this new technology have been much too powerful in Washington to date to allow support for these projects. The United States Department of Energy (DOE) has been the source of some of the most challenging obstacles to hydrogen energy programs. There is no technical field where the United States is further behind in supporting research than in the hydrogen energy field. Yet, despite efforts on the part of Congress to specifically dictate the need for a hydrogen energy program to the DOE, still the agency refuses to get serious about the technology. This problem must be dealt with in a positive way before Step 2 can make forward progress in this country. The DOE must appoint a hydrogen "Czar" within the agency, from the ranks of the thousands of scientists who really believe in the potential of the hydrogen option, and then provide him with a reasonable budget to develop and watch over a national hydrogen energy program. Fortunately, foreign nations are not plagued with this problem and have already launched serious hydrogen energy demonstration projects.

Step 3. When the technology has been seriously demonstrated and tested, then a major educational program will be needed to communicate to the public that the new hydrogen option is ready for commercialization. Unfortunately, many people, when they hear the word hydrogen, first think of either the hydrogen bomb or the Hindenburg airship disaster. These kinds of fears must be replaced by a clear understanding of the safety of metal hydride storage vessels and the other safety features of hydrogen fuel systems.

It has been my experience that when people understand the incredible advantages of hydrogen as compared to fossil fuels, they become genuinely excited about the subject and are anxious to do what they can to help out. First though, the story needs to be told. Towards this end, I anticipate the need for several independent approaches. Documentary films, colorful books, and provocative lectures will all help but will not be enough. I expect that science fiction novels, entertainment movies, and television—coupled with a strong effort by the science educators of the nation—will be required to accomplish such a mammoth communication undertaking. Once this step has been successfully carried out, then no other force nor vested interest will be able to stop the momentum. As Jules Verne once put it, "Nothing is as powerful as an idea whose time has come." People are ready for some good news. They love this planet and are genuinely concerned about what is happening. Once they learn that there is a real alternative—a perfect alternative—they will demand it!

Step 4. Finally we get to commercial implementation. This is the step where the scientists begin to hand off the technology to industry and to entrepreneurs. To induce industry to make the very large investments necessary to implement the technology, there must be a way to get a return on that investment and to make a profit. This often can best be accomplished by obtaining patents and offering to license them to industry. The most expensive implementation of a technology is almost always the first implementation. If a company has protection in the marketplace for a certain period of time, then it will be more willing to make the early investment which is required to get things started. A patent intentionally provides this type of protection and consequently an important incentive to industry.

Another method of providing incentive is through government assistance. This type of incentive might be in the form of a tax break or price guarantee. Government incentives could be very helpful to encourage companies interested in building hydrogen production facilities. The important fact is that industry will get involved only when they can make a profit, and only industry really has the muscle to make the hydrogen energy dream "wake up".

Step 5. The last step in my plan for implementing hydrogen energy is to hope that we will have completed the first four steps in time. No switch-over to a new fuel is going to take place overnight. Existing fossil fuel utilization is not going to end abruptly, but rather will be phased out over a period of time. The real world practical considerations of an undertaking of this magnitude make this fact a reality. All the more reason to get started as soon as is practical. Many of the serious problems we are creating by continuing to burn fossil fuels will take years, even hundreds of years, to undo.

I sincerely believe that at the writing of this opinion we have largely completed step one. With the very real changes to the global weather patterns, the threat of an ominous drought (which many of us in the scientific community expect in the near future), and the fact that we have already gone to war attempting to keep the important oil producing regions stable, the people of this planet are ready to stretch harder for a solution than ever before in recent history. Meanwhile, the fuel cell efficiency advantage makes hydrogen systems suddenly competitive in both cost and performance.

This incredible idea of using hydrogen as a universal fuel has finally come of age, has finally become powerful. Working together we can make it happen. I only pray we will get about it while there is still time—while there is still rain. God bless us all towards that end!

APPENDIX

HYDROGEN EQUIVALENTS

	1 m³ Hydrogen	1 ft³ Hydrogen	1 liter Liquid Hydrogen	1 gal Liquid Hydrogen	1 kilogram Hydrogen
Hydrogen Volume					
Gas					
cubic meters (STP)	1	0.0264	0.792	3.00	11.2
cubic feet (NTP)	37.9	1	30.0	114	423
Liquid					
liters (nbp)	1.26	0.0333	1	3.78	14.1
gallons (nbp)	0.334	0.00880	0.264	1	3.72
Hydrogen Mass					
kilograms	0.0896	0.00236	0.0709	0.268	1
pounds	0.198	0.00521	0.156	0.592	2.20
tons	9.87×10^{-5}	2.60×10^{-6}	7.82×10^{-5}	2.96×10^{-4}	1.10×10^{-3}
Alternative Fuels (LHV)					
Gasoline					
liters	0.352	0.00929	0.279	1.06	3.93
gallons	0.0930	0.00246	0.0737	0.279	1.04
Barrels of Crude	0.00176	4.66×10^{-5}	0.00140	0.00529	0.0197
Methanol					
liters	0.676	0.0178	0.536	2.03	7.55
gallons	0.179	0.00471	0.142	0.535	1.99
Diesel					
liters	0.279	0.00737	0.221	0.837	3.12
gallons	0.0738	0.00195	0.0584	0.221	0.824
Jet Fuel					
liters	0.287	0.00757	0.227	0.860	3.20
gallons	0.076	0.00200	0.0600	0.227	0.846
Methane (scf)	11.4	0.301	9.05	34.2	128
Propane (scf)	4.48	0.118	3.55	13.4	50.1
Butane (scf)	3.45	0.091	2.73	10.3	38.5
Coal					
Anthracite (tons)	3.97×10^{-4}	1.05×10^{-5}	3.15×10^{-4}	1.19×10^{-3}	4.44×10^{-3}
Bituminous (tons)	3.92×10^{-4}	1.04×10^{-5}	3.11×10^{-4}	1.18×10^{-3}	4.38×10^{-3}
Lignite (tons)	7.31×10^{-4}	1.93×10^{-5}	5.79×10^{-4}	2.19×10^{-3}	8.16×10^{-3}
Electricity					
kilowatt-hours	3.00	0.0791	2.38	8.99	33.5
megawatt-hours	0.003	7.91×10^{-5}	0.00238	0.00899	0.0335
Hydrogen Heat of Combustion					
Higher Heating Value					
gigajoules	0.0128	0.00034	0.0101	0.0383	0.143
10^6 Btu	0.0121	3.19×10^{-4}	0.0096	0.0363	0.135
Btu	12,100	319	9,600	36,300	135,000
kilocalories	3,100	80.5	2,400	9,100	34,100
Lower Heating Value					
gigajoules	0.0108	2.85×10^{-4}	0.0086	0.0324	0.121
10^6 Btu	0.0102	2.70×10^{-4}	0.0081	0.0307	0.114
Btu	10,200	270	8,100	30,700	114,000
kilocalories	2,600	68.0	2,040	7,700	28,800

HYDROGEN EQUIVALENTS

	1 Pound Hydrogen	1 Liter Gasoline	1 gallon Gasoline	1 kW-hr Electricity	1 gigajoule Hydrogen
Hydrogen Volume					
Gas					
cubic meters (STP)	5.04	2.84	10.8	0.333	78.3
cubic feet (NTP)	191	108	407	12.6	2,970
Liquid					
liters (nbp)	6.40	3.58	13.6	0.421	98.8
gallons (nbp)	1.69	0.947	3.58	0.111	26.1
Hydrogen Mass					
kilograms	0.454	0.256	0.967	0.0300	7.04
pounds	1	0.563	2.13	0.0661	15.5
tons	5.00×10^{-4}	2.82×10^{-4}	1.07×10^{-3}	3.31×10^{-5}	7.76×10^{-3}
Alternative Fuels (LHV)					
Gasoline					
liters	1.78	1	3.78	0.117	27.6
gallons	0.469	0.264	1	0.0310	7.28
Barrels of Crude	0.00890	0.00501	0.0190	5.88×10^{-4}	0.138
Methanol					
liters	3.41	1.92	7.27	0.225	52.9
gallons	0.901	0.507	1.92	0.0596	14.0
Diesel					
liters	1.41	0.793	3.00	0.0931	21.9
gallons	0.372	0.210	0.793	0.0246	5.78
Jet Fuel					
liters	1.45	0.815	3.08	0.0957	22.5
gallons	0.382	0.215	0.815	0.0253	5.94
Methane (scf)	57.6	32.4	123	3.81	894
Propane (scf)	22.6	12.7	48.2	1.50	351
Butane (scf)	17.4	9.79	37.0	1.15	270
Coal					
Anthracite (tons)	2.00×10^{-3}	1.13×10^{-3}	4.27×10^{-3}	1.32×10^{-4}	3.11×10^{-2}
Bituminous (tons)	1.98×10^{-3}	1.11×10^{-3}	4.22×10^{-3}	1.31×10^{-4}	3.07×10^{-2}
Lignite (tons)	3.69×10^{-3}	2.08×10^{-3}	7.86×10^{-3}	2.44×10^{-4}	5.72×10^{-2}
Electricity					
kilowatt-hours	15.1	8.52	32.2	1	235
megawatt-hours	0.0151	0.00852	0.0322	0.001	0.235
Hydrogen Heat of Combustion					
Higher Heating Value					
gigajoules	0.0644	0.0363	0.137	0.00426	1
10^6 Btu	0.0610	0.0344	0.130	0.00404	0.948
Btu	61,000	34,400	130,000	4,040	948,000
kilocalories	15,400	8,660	32,800	1,020	239,000
Lower Heating Value					
gigajoules	0.0544	0.0307	0.116	0.00360	0.845
10^6 Btu	0.0516	0.0291	0.110	0.00341	0.801
Btu	51,600	29,100	110,000	3,410	801,000
kilocalories	13,000	7,300	27,700	860	202,000

UNIT CONVERSIONS

Mass

	1 g	1 kg	1 metric ton	1 lb$_m$	1 ton
grams (g)	1	1000	1×10^6	453.592	907,185
kilograms (kg)	0.001	1	1000	0.453592	907.185
metric tons	1×10^{-6}	0.001	1	4.5359×10^{-4}	0.907185
ounces (oz.)	0.035274	35.274	35,274	16	32,000
pounds (lb$_m$)	2.2046×10^{-3}	2.20462	2204.62	1	2000
tons	1.1023×10^{-6}	1.1023×10^{-3}	1.1023	5×10^{-4}	1

Length

	1 cm	1 m	1 km	1 in.	1 ft
millimeters (mm)	10	1000	1×10^6	25.4	304.8
centimeters (cm)	1	100	1×10^5	2.54	30.48
meters (m)	0.01	1	1000	0.0254	0.3048
kilometers (km)	1×10^{-5}	0.001	1	2.54×10^{-5}	3.048×10^{-4}
inches (in.)	0.3937	39.37	39,370	1	12
feet (ft)	0.032808	3.2808	3280.8	0.083333	1
yards (yd)	0.010936	1.0936	1093.6	0.027778	0.333333
miles (mi.)	6.214×10^{-6}	6.214×10^{-4}	0.6214	1.5783×10^{-5}	1.8939×10^{-4}

Volume

	1 cm^3	1 m^3	1 liter	1 ft^3	1 gal
cubic centimeters (cm^3)	1	1×10^6	1000	28,317	3785.4
cubic meters (m^3)	1×10^{-6}	1	0.001	0.028317	3.7854×10^{-3}
milliliters (ml)	1	1×10^6	1000	28,317	3785.4
liters (l)	0.001	1000	1	28.317	3.7854
cubic inches (in.^{3})	0.0610237	61,023.74	61.02374	1728	231
cubic feet (ft^3)	3.5315×10^{-5}	35.3147	0.0353147	1	0.133681
quarts (qt)	1.0567×10^{-3}	1056.69	1.056688	2.99221	4
imperial gallons	2.1997×10^{-4}	219.97	0.21997	6.22884	0.8327
gallons (gal)	2.6417×10^{-4}	264.172	0.264172	7.4805	1

Force

	1 N	1 dyn	1 lb$_f$
newtons (N)	1	1×10^{-5}	4.4482
dynes (dyn)	1×10^5	1	444,820
pounds (lb$_f$)	0.22481	2.2481×10^{-6}	1

UNIT CONVERSIONS

Pressure

	1 Pa	1 bar	1 atm	1 mm Hg	1 psi
pascals (Pa)	1	1×10^5	101,325	133.3224	6894.76
newtons/meter^2 (N/m^2)	1	1×10^5	101,325	133.3224	6894.76
dynes/centimeter^2 (dyn/cm^2)	10	1×10^6	1,013,250	1333.224	68,947.6
bars	1×10^{-5}	1	1.01325	1.3332×10^{-3}	0.068948
atmospheres (atm)	9.8692×10^{-6}	0.986923	1	1.3158×10^{-3}	0.068046
millimeters mercury (mm Hg)	7.5006×10^{-3}	750.062	760	1	51.7149
torr	7.5006×10^{-3}	750.062	760	1	51.7149
inches mercury (in. Hg)	2.953×10^{-4}	29.53	29.9213	0.03937	2.036
pounds/inch^2 (psi)	1.4504×10^{-4}	14.5038	14.696	0.019337	1

Energy

	1 J	1 GJ	1 kW-hr	1 cal	1 Btu
joules (J)	1	1×10^9	3.6×10^6	4.184	1054.35
gigajoules (GJ)	1×10^{-9}	1	0.0036	4.184×10^{-9}	1.054×10^{-6}
newton-meters (N-m)	1	1×10^9	3.6×10^6	4.184	1054.35
ergs	1×10^7	1×10^{16}	3.6×10^{13}	4.184×10^7	1.054×10^{10}
dyne-centimeters (dyn-cm)	1×10^7	1×10^{16}	3.6×10^{13}	4.184×10^7	1.054×10^{10}
kilowatt-hour (kW-hr)	2.778×10^{-7}	277.8	1	1.1622×10^{-6}	2.9287×10^{-4}
megawatt-hours (MW-hr)	2.778×10^{-10}	0.2778	0.001	1.1622×10^{-9}	2.9287×10^{-7}
calories (cal)	0.23901	2.390×10^8	860,420	1	251.996
kilocalories (kcal)	2.3901×10^{-4}	239,010	860.42	0.001	0.251996
foot-pounds (ft-lb$_f$)	0.7376	7.376×10^8	2.6552×10^6	3.08596	777.65
British thermal units (Btu)	9.48×10^{-4}	948,000	3414.52	3.9683×10^{-3}	1
MM Btu (10^6 Btu)	9.48×10^{-10}	0.948	3.41452×10^{-3}	3.9683×10^{-9}	1×10^{-6}

Power

	1 W	1 cal/s	1 ft-lb$_f$/s	1 Btu/s	1 hp
watts (W)	1	4.184	1.356	1054.35	745.700
kilowatts (kW)	0.001	0.004184	1.356×10^{-3}	1.05435	0.745700
megawatts (MW)	1×10^{-6}	4.184×10^{-6}	1.356×10^{-6}	1.0544×10^{-3}	7.4570×10^{-4}
joules/second (J/s)	1	4.184	1.356	1054.35	745.700
calories/second (cal/s)	0.23901	1	0.324	251.996	178.227
foot-pounds/second (ft-lb$_f$/s)	0.7376	3.08596	1	777.65	550
Btu/second (Btu/s)	9.48×10^{-4}	3.9683×10^{-3}	1.286×10^{-3}	1	0.7073
horsepower (hp)	1.341×10^{-3}	5.611×10^{-3}	1.818×10^{-3}	1.4139	1

BIBLIOGRAPHY

Adt, R.R., Herschberger, D.L., Kartage, T. and Swain, M.R. 1973. "The Hydrogen-Air Fueled Automobile Engine (Part 1)." Proc. 8th Intersociety Energy Conversion Engineering Conference, University of Pennsylvania, Philadelphia, p. 194, 197, August.

Adt, R.R., Greenwall, H., and Swain, M.R. 1974. "The Hydrogen/Methanol-Air Breathing Automobile Engine." *Proceedings, The Hydrogen Economy Miami Energy (THEME) Conference*, pp. S10-37-48, Miami Beach, Florida, March.

Amphlett, J.C., Farahani, M., Mann, R.F., Peppley, B.A., and Roberge, P.R. 1991. "Operating Characteristics of a Solid Polymer Fuel Cell." *Proceedings--Project Hydrogen '91 World Conference*, Independence, Missouri: American Academy of Science.

Anderson, J.H. and Anderson, J.H. Jr. 1966. "Thermal Power from Sea Water." *Mechanical Engineering*, Vol. 88, N.4, April.

Anzilotti, W.F., and Tomsie, V.J. 1954. "Combustion of Hydrogen and Carbon Monoxide as Related to Knock." *The Combustion Institute, Fifth Syposium on Combustion*.

Arnett, R.W., Mullen, L.O. and Warren, K.A. 1960. "Some Methods for Reducing Heat Leak Through Support Members in Liquefied Gas Storage Vessels." *Advances in Cryogenic Engineering*, Vol. 4, p. 410, New York: Plenum Press.

Austin, A.L., Higgins, G.H., and Howard, J.H. 1973. "The Total Flow Concept for Recovery of Energy from Geothermal Hot Brine Deposits."

Austin, L.G. 1967. "Fuel Cells," NASA SP120.

Auttal, L.J. et al. 1974. "Hydrogen Generation by Solid Polymer Electrolyte Water Electrolysis." General Electric Company, Lynn, Massachusetts.

Bain, A.L. and Boggs, W.H. 1975. "NASA Space Program Experience in Hydrogen Transportation and Handling." Paper presented at the Hydrogen Energy Fundamentals Symposium, University of Miami, Coral Gables, Florida, March 3-5.

Bainbridge, R. and Horton, T.R. 1971."The Production of Liquid Hydrogen at the Rocket Propulsion Establishment." Technical Report No. 71/17, Rocket Propulsion Establishment, Wescott, U.K., December.

Baker, N.R. 1974. "Oxides of Nitrogen Control Techniques for Appliance Conversion to Hydrogen Fuel," Proceedings 9th Intersociety Energy Conversion Engineering Conferences, San Francisco, California, August 26-30.

Balje, O.E. 1956. "Drag Turbine Performance." ASME paper No. 56-AV-6, March.

Baranov, V.I., Muchnik, C.F., and Trushevskii, S.N. 1969. "Investigation of High Temperature Solar Energy Absorbers and Thermal Storage Devices." *Semiconductor Solar Energy Converters*, V.A. Baum, Editor, Consultant's Bureau, New York, p. 183.

Bardos, Russel. 1977. "Gasifiers In Industry Program." Presented at the Forth Annual International Conference of Coal Gasification, Liquefaction and Conversion to Electricity, University of Pittsburg, August.

Bartlit, J.R. 1975. "Liquid Hydrogen Handling, Transport, and Storage." *Proceedings of the Cornell International Symposium and Workshop on the Hydrogen Economy*, p. 95.

Baum, W.A. 1955. "Meteorology and Utilization of Solar Energy." *Solar Energy Research*, F. Daniels and J.A. Duffie, Editors, University of Wisconsin Press, p. 15.

Becker, J.V. 1970. Paper presented at the 7th Congress of the International Council of the Aeronautical Sciences, Rome, Italy, Sept.

Becker, J.V. 1971. *Astronautics and Aeronautics*, **9**, No. 8, 32.

Benjamin, T.G., Laurens, R.M. and Camara, E.H. 1991. "Molten Carbonate Fuel Cell Development at M-C Power." *Proceedings--Project Hydrogen '91 World Conference*, Independence, Missouri: American Academy of Science.

Bergman, P.D., Plants, K.D., Demeter, J.J., and Bienstock, D. 1973. "An Economic Evaluation of MHD-Steam Powerplants Employing Coal Gasification." BM-RI-7796, Bureau of Mines, U.S. Department of Interior.

Beyer, R.B. and Woolley, R.L. 1977. "Refueling Hydrogen Transit Fleets Part A - Economics." Fourth International Symposium on Automotive Propulsion Systems sponsored by the NATO Committee on the Challenges of Modern Society, Washington, D.C.

Billings, R.E., Sanchez, M, Cherry, P., Eyre, D.B. 1991. "LaserCel1 Prototype Vehicle." *Proceedings--Project Hydrogen '91 World Conference*, Independence, Missouri: American Academy of Science.

Billings, R.E. 1975. "Hydrogen's Potential As A Vehicular Fuel for Transportation." 759174, Tenth IECEC, Delaware, pp. 1165-1175.

Billings, R.E. 1966. "The Hydrogen Engine: A Solution to Pollution." Report distributed at the 17th International Science Fair, Dallas, Texas, May.

Billings, Roger E. 1976. "A HydrogenPowered Mass Transit System." Proceedings 1st World Hydrogen Energy Conference, 7C, 27-42, Miami Beach, March 1-3.

Billings, Roger E. 1976. "Hydrogen Storage in Automobiles Using Cryogenics and Metal Hydrides." Proceedings 1st World Hydrogen Energy Conference, Miami Beach, March 1-3.

Billings, R.E., and Lynch, F.E. 1973. "History of Hydrogen-Fueled Internal Combustion Engines." Billings Tech Library, 73001, American Academy of Science, Independence, Missouri.

Billings, R.E. 1978. "Hydrogen Homestead." Billings Tech Library 78005, American Academy of Science, Independence, Missouri.

Billings, Roger E. 1976. "A Comparison of Operational Economics of Transportation Vehicles Operated on Gasoline and Coal-Generated Hydrogen." Proceedings American Chemical Society National Meeting, New York City, April 4-9.

Billings, R.E., and Lynch, F.E. 1973. "Performance and Nitric Oxide Control Parameters of the Hydrogen Engine." Billings Tech Library 73002, American Academy of Science, Independence, Missouri.

Billings, R.B. 1976. "Survey of Hydrogen Energy Application Projects." Billings Tech Library 76009, American Academy of Science, Independence, Missouri.

Blackwell, D.D. 1974. "Geothermal Resources of Marysville, Montana." Proceedings, 9th IECEC Conference, San Francisco, California, August 26-30.

Blankenship, D.T. and Winget, G.D. 1973. "Hydrogen Fuel: Production by Bioconversion." 8th Intersociety Energy Conversion Engineering Conference, University of Pennsylvania, Philadelphia, Pennsylvania, August.

Bockris, J.O.'M., Bonciocat, N., and Gutmann, F. 1974. *An Introduction to Electrochemical Science*, London: Wydeham Press, p. 78.

Bockris, J.O.'M., and Drazie, D. 1972. *Electrochemical Science*, London: Taylor and Francis.

Bockris, J.O.'M, Mackenzie, J.D., and White, J.I. 1960. *Physicochemical Measurements at High Temperatures*, London: Butterworths, p. 95-97.

Bockris, J.O.'M., and Srinivasan, S. 1970. *Fuel Cells: Their Electrochemistry*, New York: McGraw-Hill.

Bockris, J.O.'M. and Reddy, A.K.N. 1973. *Modern Electrochemistry*, Plenum: Rosetta Edition, New York.

Bockris, J.O.'M., and Reddy, A.K.N. 1973. *Modern Electrochemistry*, New York: Rosetta Edition; Plenum, p. 1132.

Breshears, R., Cotrill, H. and Rupe, J. 1973. "Partial Hydrogen Injection into Internal Combustion Engines - Effect on Emissions and Fuel Economy." EPA-First Symposium on Low Pollution Power Systems Development, Ann Arbor, Michigan, Oct. 14-19.

Breshears, R., Cotrill, H. and Rupe, J. 1974. "Hydrogen Injection for Internal Combustion Engines." presented at EPA Alternative Automotive Power Systems Coordination Meeting, Ann Arbor, Michigan, May.

Broeker, R.J., Bress, D.F. 1977. "Means to Solve an Energy Problem." *Heat Engineering*, Foster Wheeler Corporation, Livingston, New Jersey, September.

Buchner, H. and Saufferer, H. 1974. "The Short and Medium Term Development and Practical Application of Hydrogen Powered Vehicles." Presented at NATO Conference, Dusseldorf.

Burger, James M., and et al. 1974. "Energy Storage for Utilities via Hydrogen Systems." Proceedings 9th Intersociety Energy Conversion Engineering Conferences, San Francisco, California, August 26-30.

Burke, J.C., Byrnes, W.R., Post, A.H. and Ruccia, F.E. 1960. "Pressurized Cooldown of Cryogenic Transfer Lines." *Advances in Cryogenic Engineering*, Vol. 4, p. 378, New York: Plenum Press.

Burstall, A.F. 1927. "Experiments on the Behavior of Various Fuels in a High Speed Internal Combustion Engine." *Institute of Automotive Engineers*, Vol. 22, December.

Burstall, A.F. 1927. *Proceedings Inst. Automobile Engineers*, Vol. 22, p. 358.

Burton, J.D. 1966. "The Prediction and Improvement of Regenerative Turbo-Machine Performance." Thesis, Southampton University.

Burton, D.W. 1963. "Review of Regenerative Compressor Theory." AEC TID 7631, *Rotating Machinery for Gas-Cooled Reactor Application*, pp. 228-242.

Cameron, H.M. and Morgan, N.E.1964. "Development of Hydrogen-Oxygen Fueled 3-Kilowatt Internal Combustion Engine." AIAA Paper No. A64-756, September.

Campbell, B.C. 1976. "Development of a Low Capital Cost Electrolyzer." 1st World Hydrogen Energy Conference, Miami Beach, Florida.

Campbell, B.C. 1978. "Development of Billings SPE Electrolyzer." 2nd World Hydrogen Energy Conference, Zurich, Switzerland, August: Billings Tech Library 78004, American Academy of Science, Independence, Missouri.

Campbell, B.C. 1978. "Hydrogen Economy: An Alternative." Billings Tech Library 78007, American Academy of Science, Independence, Missouri.

Chemical and Engineering News, 12 November 1973, p. 11.

Cherry, W.R. 1972. "The Generation of Pollution-Free Electrical Power from Solar Energy." *Journal of Engineering for Power*, p. 78, April.

Chlumsky, V. 1965. "Reciprocating and Rotary Compressors." E.&F.N. Spon Std., London.

Cole, D.E. 1973. "Update on the Wankel Rotary Engine." EPA-First Symposium on Low Pollution Power Systems Development, Ann Arbor, Michigan, October 14-19.

Commander, J.C. and Ratter, L. 1965. "An Economic Analysis of Perlite Versus Superinsulation in Liquid Hydrogen Storage Vessels." NASA CR-54720, October.

Coward, H.F., and Jenks, G.W. 1952. "Limits of Flammability of Gases and Vapors," U.S. Bureau of Mines, Bulletin 503.

Cox, K.E. and Williamsen, K.D. Editors. 1975. *Hydrogen: Its Technology and Implications*, Vols. I-II: CRC Press, Cleveland, Ohio.

Crewdson, E. 1956. "Water-Ring Self-Priming Pumps." Proceedings of Institution of Mechanical Engineers, Vol. 170, n.13, pp. 407-417.

Damjanovic, A., Sepa, D. and Bockris, J.O.'M. 1968. "J. Res. Inst. Catalysis." Hokkaido Unic., 16.

deBeni, G. and Marchetti, C. 1972. "Mark-1, A Chemical Process to Decompose Water Using Nuclear Heat." Paper presented to the American Chemical Society Meeting, Boston, April.

Deen. J.I. and Schoeppel, R.J. 1970. Paper presented at the Frontiers of Power Technology Conference, Stillwater, Oklahoma, October.

Deyou, B. 1991. "Research and Development Program Related in Mitigation of Global Warming Trends in China." *Proceedings--Project Hydrogen '91 World Conference*, Independence, Missouri: American Academy of Science.

Downs, D., Walsh, A.D., and Wheeler, R.S. 1951. "Knock in the Spark-Ignition Engine." *Trans. Roy. Soc. (London)* 243, #870, July 19.

Drell, I.L. and Belles, F.E. 1958. "Survey of Hydrogen Combustion Properties," NASA 1383.

Dupree, W.G., Jr. and West, J.A. 1972. "United States Energy Through the Year 2000." A U.S. Department of the Interior Report, December.

Eccleston, D.B. and Fleming, R.D. 1972. "Clean Automotive Fuel: Engine Emissions Using Natural Gas, Hydrogen-Enriched Natural Gas, and Gas Manufactured from Coal (Synthane)." Tech. Prog. Report-48, U.S. Bureau of Mines, Bartlesville, Oklahoma, February.

Eckert, R.G. and Drake, R.M. 1969. *Heat and Mass Transfer*, New York: McGraw-Hill Book Company, Inc.

Egerton, A., Smith, F.L, Libbelohde, A.R. 1935. "Estimation of the Combustion Products from the Cylinder of the Petrol Engine and Its Relation to 'Knock'." *Phil. Trans., Roy. Soc. of London*, Vol. 234 A, p. 466.

Elliott, M.A. and Turner, N.C. 1972. Presented at the American Chemical Society Meeting, Boston Massachusetts, April.

Energy Profile, February 1974, p. 23.

Erren, R.A., and Campbell, W.H. 1933. "Hydrogen: A Commercial Fuel for Internal Combustion Engines and Other Purposes." *Journal of the Institute of Fuel*, Vol. 6, No. 29, June.

Erren, R.A. 1936. "Internal Combustion Engine Using Hydrogen as Fuel." U.S. Patent No. 2,183,674. Application Sept. 10, 1936.

Erren, R.A. 1930. "Method for Driving Internal Combustion Engines." U.S. Patent No. 1,901,709. Filed Mar. 15, 1930.

Erren, R.A. 1938. "Method of Charging Internal Combustion Engines." U.S. Patent No. 2,164,234. Application Sept. 17, 1938.

Escher, W.J.D. 1972. "On the Higher Energy Form of Water (H_2O') in Automotive Vehicle Advanced Power Systems." Proceedings 7th Intersociety Energy Conversion Engineering Conference, San Diego, California, September.

Escher, W.J.D. 1975. "Hydrogen Fueled Internal Combustion Engine, A Technical Survey of Contemporary U.S. Projects." United States Energy Research & Development Administration, No. PR-51.

Escher, W.J.D. and Brewer, G.D. 1974. Paper presented at the AIAA 12th Aerospace Sciences Meeting, Washington, D.C., February.

Escher, W.J.D. 1972. "Helio-Poseidon, A Macro System for the Production of Storable, Transportable Energy from the Sun and the Sea." Escher Technology Associates Report PR-8, February.

Escher, W.J.D. 1973. "Prospects for Liquid Hydrogen Fueled Commercial Aircraft." Report PR-37, Escher Technology Associates, St. Johns, Michigan, September.

Feirtag, M. 1972. "Cars in Search of the Future." *Technology Review*, October/November.

Flieder, W.G., Smith, W.J. and Wetmore, K.R. 1960. "Flexibillity Considerations for the Design of Cryogenic Transfer Lines." *Advances in Cryogenic Engineering*, Vol. 5, p. 111, New York: Plenum Press.

Frass, A.P. 1948. *Combustion Engines*, New York, N.Y.: McGraw-Hill Book Company, Inc.

Funk, J.E. and Reinstrom, R. 1966. *I & E.C. Process Design Development*, 5, p. 336, July.

Funk, J.E., Conger, W.I. and Carty, R.H. 1974. Paper presented at the Hydrogen Economy Miami Energy (THEME) Conference, March.

Funk, J.E. 1972. Paper presented at the Am. Chem. Soc. Div. Fuel Chem., p. 79, April.

Furuhama, S. 1991. "Hydrogen Engine Technology." *Proceedings--Project Hydrogen '91 World Conference*, Independence, Missouri: American Academy of Science.

Gahimer, G., et al. 1976. "Experimental Demonstration of an Iron Chloride Thermochemical Cycle for Hydrogen Production." Eleventh Intersociety Energy Conversion Engineering Conference, State Line, Nevada.

Gallone, P. 1972. "Water Electrolysis." *Encyclopedia of Electrochemistry*, Robert E. Krieger Publishing Company, Huntington, New York.

Gates, P.S. 1964. "Peripheral-Compressor Performance on Gases with Molecular Weights of 4 to 400." American Society of Mechanical Engineers paper 64 WA/FE-25, for meeting November 29-December 4.

Gerrish, H.C. and Foster, H.H. 1935. "Hydrogen as an Auxiliary Fuel in Compression-Ignition Engines." Report No. 535, National Advisory Committee for Aeronautics (NACA).

Ghergel, Magda, Melinte, Sofia, 1991. "Researches Concerning the Photovoltaic Effects and Some Physical Interface Processes at Mixed Semiconductor Heterostructures with Photosynthetic Pigment." *Proceedings--Project Hydrogen '91 World Conference*, Independence, Missouri: American Academy of Science.

Gibbs, C.W., Editor, 1969. "Compressed Air and Gas Data." Ingersoll-Rand Company, New York.

Gorbell, B.N., and Prater, K.B. 1991. "The Ballard/BC Bus Demonstration Program." *Proceedings--Project Hydrogen '91 World Conference*, Independence, Missouri: American Academy of Science.

Gordon, S. and McBride, B.J. 1971. "Computer Program for Calculation of Complex Chemical Equilibrium Compositions, Rocket Performance, Incident and Reflected Shocks, and Chapman-Jouguet Detonations." NASA SP273.

Gough, G.H. 1973. Paper presented at the Cornell National Symposium on the Hydrogen Economy, August.

Grabow, G. 1966. "Investigation on Peripheral Pumps." Second Conference on Flow Machines, Budapest, pp. 147-166, October.

Gray, H.R. 1972. "Hydrogen Environment Embrittlement." NASA TM X-68088, Lewis Research Center, Cleveland, Ohio.

Gregory, D.P. 1973. "A Hydrogen Energy System." American Gas Association, Chicago, Illinois.

Gregory, D. assisted by Anderson, P.J., Dufour, R.J., Elkins, R.H., Escher, W.J.D., Foster, R.B., Long, G.M., Wurm, J. and Yie, G.G. 1973. "The Hydrogen-Energy System." Prepared for American Gas Association by I.G.T., p. 111-3.

Gregory, D.P., and Dufour, R.J. 1972. "Utilization of Synthetic Fuels Other than Hydrogen." Institute of Gas Technology Report on Project 8936 for ORNL, May.

Gregory, D.P. and Pangborn, J.B. 1976. "Hydrogen Energy." *Annual Reviews of Energy*, Vol. 1, Annual Reviews, Inc., Palo Alto, California.

Grigger, J.C. 1972. "Lead Dioxide Anode." *Encyclopedia of Electrochemistry*, Robert E. Krieger Publishing Company, Huntington, New York.

Grot, W.G.F., et al. 1972. "Perfluorinated Ion Exchange Membranes." Plastics Department, E.I. duPont de Nemours and Company, Wilmington, Delaware.

Grot, W.G.F., Munn, G.E., Walmsley, P.N. 1972. "Perfluorinated ION Exchange Membranes." Plastics Department, E.I. duPont de Nemours & Company, Wilmington, Delaware.

Gupta, A.S. and Thodos, G. 1964. "Transitional Behavior for the Simultaneous Mass and Heat Transfer of Gases Flowing Through Packed and Distended Beds of Spheres." *Ind. Engineering Chemistry Fundamentals* 3, 218-220.

Gupta, A.S. and Thodos, G. 1963. "Direct Analogy Between Mass and Heat Transfer to Beds of Spheres." *Journal American Institute of Chemical Engineers* 9, 751-754.

Hadden, L.D. 1978. "The Economics of Producing Hydrogen From a Small-Air Blown Coal Gasifier." Billings Tech Library 78003, American Academy of Science, Independence, Missouri.

Haettinger, G.C. 1966. "Considerations in the Desing, Selection and Use of Vacuum Insulated Pipe." *Advances in Cryogenic Engineering*, Vol. 11, p. 98, New York: Plenum Press.

Haldane, J.B.S. 1923. "Daedalus, or Science and the Future." A paper read to the Heretics on February 4, 1923, Cambridge, Mass.

Hallett, N.C. 1974. "Cost, Study and Systems Analysis of Siquid Hydrogen Production." Air Products and Chemicals Inc. Contract NAS 23894, June.

Hammond, A.L., Metz, W.D., and Maugh II, T.H. 1973. "Energy and the Future." American Association for the Advancement of Science, Washington, D.C.

Hansel, James G., Mattern, Glenn W., Miller, Robert N. 1991. "Safety Considerations in the Design of Hydrogen-Powered Vehicles." *Proceedings--Project Hydrogen '91 World Conference*, Independence, Missouri: American Academy of Science.

Harral, J.K.A., Jones, M.R., and Hall, D.E. 1977. "A Comparison of Energy Options Gas or Electricity." Pacific Gas and Electric Company, San Francisco, California.

Harris J.A. and Van Wanderham, M.C. 1973. Pratt and Whitney Aircraft Report NASA CR-124394, July.

Hattingen, U. and Jordan, W. 1974. "Wasserstoff als Zusatzkraft." *VDI Nachrichten*, No. 21/24, pp. 4-5, May.

Hausz, W., Leeth, G., Lueck, D., and Meyer, C. 1972. "Hydrogen Systems for Electric Energy." 72TMP-15, General ElectricCompany.

Hausz, W., Leeth, G., and Meyer, C. "Eco-Energy." General Electric TEMPO, Report 729206.

Heitbrink, E.H., McBreen, J., Selis, S.M., Tricklebank, S.B., and Witherspoon, R.R. 1972. *The Electrochemistry of Cleaner Environments*, ed. J.O.'M. Bockris, New York: Plenum Press, p. 47.

Helmore, W. and Stokes, P.H. 1930. "Hydrogen-Cum-Oil Gas as an Auxiliary Fuel for Airship Compression Ignition Engines." Report No. E3219 Brit. Roy. Aircraft. Establ., October.

Hendriksen, D.L. and et al. 1976. "Prototype Hydrogen Automobile Using Metal Hydride" Proceedings 1st World Hydrogen Energy Conference, Miami Beach, March 1-3.

Heronemus, W.E. 1972. "The United States Energy Crisis: Some Proposed Gentile Solutions." ASME/IEEE Joint Meeting, West Springfield, Massachusetts, January 12.

Heywood, J.B. and Womack, G.J. 1969. *Open-Cycle MHD Power Generation*: Pergamon Press, pp. 212-217, 240-241, 781-784.

Hodara, I. 1991. "Solar Fuels and Chemicals Research at the Weizmann Institute of Science." *Proceedings--Project Hydrogen '91 World Conference*, Independence, Missouri: American Academy of Science.

Hodara, I. 1991. "Solar Fuels andChemicals Research at the Weizmann Institute of Science." *Proceedings--Project Hydrogen '91 World Conference*, Independence, Missouri: American Academy of Science.

Hodara, I., Kogan, A., Gorodnev, A. 1991. "Direct Solar Water Splitting Experiment." *Proceedings--Project Hydrogen '91 World Conference*, Independence, Missouri: American Academy of Science.

Hoehn, F.W., and Dowdy, M. W. 1974. "Feasibility Demonstration of a Road Vehicle Fueled with Hydrogen-Enriched Gasoline." Presented at the 9th Intersociety Energy Conversion Engineering Conferences, San Francisco, California, August 26-30.

Hoffman, K.C., Reilly, J.J., Salzano, F.J., Waide, C.H., Wiswall, R.H., and Winsche, W.E. 1976. "Metal Hydride Storage for Mobile and Stationary Applications." SAE paper 760569, SAE Fuels and Lubricants Meeting, S. Louis, June.

Hoffman, K.C., Editor, 1973. "Energy Systems Analysis and Technology Assessment Programs." Annual Report FY 1973, BNL-17959, June.

Hoffman, K.C. et al. 1969. "Metal Hydrides as a Source of Fuel for Vehicular Propulsion." Paper SAE 690232 presented at the International Automotive Engineering Conference, Detroit, January 13-17.

Hoffman, K.C. 1972. "The U.S. Energy System--A Unified Planning Framework." Doctoral dissertation, Polytechnic Institute of Brooklyn, June.

Hoffman, K.C., and et al. 1976. "Metal Hydride Storage for Mobile and Stationary Applications." Society of Automotive Engineers, Inc. No. 60569.

Hollenberg, J.W. 1966. "Vortex Flow Study." The Singer Company, Corporate Research Report No. CR-97, (Proprietary).

Hollenberg, J.W. 1967. "Vortex Blower Development." The Singer Company, Corporate Research Report No. CR-146, July, (Proprietary).

Hord, J. (Editor) 1975. "Selected Topics on Hydrogen Fuel." Report NBS IR-75-803, U.S. National Bureau of Standards, January.

Ishaghoff, I. and Canty, J.M. 1964. "Quilted Insulation." *Advances in Cryogenic Engineering*, Vol. 9, p. 45, New York: Plenum Press.

Iverson, H.W. 1955. "Performance of the Periphery Pump." ASME Pager 53-A-102, *Transactions of the ASME*, January, pp. 1303-1316.

Jacobs, R.B., Richards, R.J. and Schwartz, S.B. 1960. "The Transfer of Liquefied Gases." *Advances in Cryogenic Engineering*, Vol. 1, p. 87, New York: Plenum Press.

Jewett, R.P., et al. 1973. "Hydrogen Environment Embrittlement of Metals." NASA CR-2163, March.

Johnson, J.E. 1973. Paper presented at the Cryogenic Engineering Conference, Atlanta, Georgia, August.

Jones, J.E., et al. 1976. "Assessment of Very High Temperature Reactors in Process Applications." Eleventh Intersociety Energy Conversion Engineering Conference, State Line, Nevada.

Just, J.S. 1944. "Hydrogen as a Substitute Fuel." *Gas and Oil Power*, Annual Tech. Review.

Karim, G.A. and Taylor, M.E. 1973. "Hydrogen as a Fuel and the Feasibility of a Hydrogen-Oxygen Engine." SAE Paper No. 730089, January.

Kelly, J.H. and Laumann, E.A. 1975. "Hydrogen Tomorrow Demands and Technology Assessment." Jet Propulsion Laboratory, Pasadena, California.

Kincaid, W. 1973. Paper presented at the Conference on the Hydrogen Economy, Cornell.

King, R.O., et al. 1958. "The Hydrogen Engine: Combustion Knock and Related Flame Velocity." *Trans. E.I.C.*, Vol. 2.

King, R.O., Wallace, W.A., and Mahapatra, B. 1948. *Can. J. of Res.*, **26**, Sec. F, 264.

King, R.O., et al. 1948. "The Hydrogen Engine and the Nuclear Theory of Ignition." *The Oxidation, Ignition and Detonation of Fuel, Vapors and Gases. Can. Jour. of Res.*, Vol. 26, Sec. F.

Kirkham, F.S. and Friver C. 1973. Paper presented at AIAA 5th Aircraft Design, Flight Test and Operations Meeting; St. Louis, Missouri, August.

Korycinski, Peter F. 1976. "Some Early Perspectives on Ground Requirements of Liquid Hydrogen Air Transports." Proceedings 1st World Hydrogen Energy Conference 5C, 33-56, Miami Beach, Florida, March 1-3.

Korycinski, Peter F. and Snow, Daniel B. 1974. "Hydrogen for the Subsonic Transport." Proc. The Hydrogen Economy Miami Energy (Theme) Conference, XII, 1-24, March 18-20.

Kreith, F. and Dean, J.W. 1968. "Cooldown and Warmup of Large Powder-Insulated Dewars." *Advances in Cryogenic Engineering*, Vol. 8, p. 536, New York: Plenum Press.

Kropschot, R.H., et al. 1960. "Multilayer Insulation." *Advances in Cryogenic Engineering*, Vol. 5, p. 189, New York: Plenum Press.

Lambe, Dr. S.M., Watson, Dr. H.C. 1991. "Hydrogen Diesel and S.I. Engine Performance Targets for Other Engines to Match." *Proceedings--Project Hydrogen '91 World Conference*, Independence, Missouri: American Academy of Science.

Laquer, H.L. 1960. "Handling Liquid Hydrogen on a Pilot-Plant Scale." *Advances in Cryogenic Engineering*, Vol. 5, p. 85, New York: Plenum Press.

Laws, J.S., Frick, V., and McConnell, J. 1969. "Hydrogen Gas Pressure Vessel Problems in the M-1 Facilities." NASA CR-1305, Washington, D.C., March.

Lee, B.S. 1970. Presented at the American Power Conference, Chicago, Illinois, April.

Lehman, P.A., Chamberlin, C.E. 1991."A Photovoltaic-Hydrogen-Fuel Cell Energy System: Control Strategy, Monitoring Design, and Preliminary Results." *Proceedings--Project Hydrogen '91 World Conference*, Independence, Missouri: American Academy of Science.

Lessing, L. 1972. *Fortune*, November, p. 210.

Lessing, L. 1973. *Chemistry in the Environment*, San Francisco: Freeman, p. 159.

Lewis Laboratory Staff, 1957. "Hydrogen for Turbojet and Ramjet Powered Flight." NACA RM E57D23, April.

Libby, W.P. 1971. *Science*, **171**, 499.

Lichty, L.C. 1967. *Combustion Engine Processes*, New York, N.Y.: McGraw-Hill Book Company, Inc.

Liebenberg, D.H., Stokes, R.W. and Edeskuty, F.J. 1960. "Chilldown and Storage Losses of Large Liquid Hydrogen Storage Dewars." *Advances in Cryogenic Engineering*, Vol. 11, p. 554, New York: Plenum Press.

Liebermann, P. 1960. "E.R.E.T.S. LOX Losses and Preventative Measures." *Advances in Cryogenic Engineering*, Vol. 2, p. 225, New York: Plenum Press.

Linde Division of Union Carbide Corporation, 1975. "Survey Study of the Efficiency and Economics of Hydrogen Liquefaction." Contract NAS 1-13395, NASA, Hampton, Virginia, April 8.

Lof, G.O.G., et al. 1946. "Solar Energy Utilization for House Heating." Report No. PB 25375, Department of Commerce, Washington, D.C.

Lotker, Michael, 1974. "Hydrogen for the Electric Utilities - Long Range Possibilities," Proc. the 9th Intersociety Energy Conversion Engineering Conferences, San Francisco, California, August 26-30.

Lucke, C.E. 1905. *Gas Engine Designs*, New York: C. Van Nostrand Company, Inc.

Lynch, Frank E. 1974. "Backfire Control Techniques for Hydrogen Fueled Internal Combustion Engines." Proceedings The Hydrogen Economy Miami Energy (Theme) Conference, S10, 27-36, March 18-20.

Mackay, Dr. D.B. 1976. "Economy of Hydrogen-Fueled Automobile Engines." Billings Tech Library 76008, American Academy of Science, Independence, Missouri.

Manabe, S. and Wetherald, R.T. 1967. *J. Atm. Sci.*, **24**, 241.

Mandelis, Andreas, Christofides, Constantinos, 1991. "Photopyroelectric PD-PVDF Solid-State Hydrogen Sensor." *Proceedings--Project Hydrogen '91 World Conference*, Independence, Missouri: American Academy of Science.

Marchetti, C. 1973. *Chemical Economy & Engineering Review*, 5,7.

Martinelli, R.C. and Nelson, D.B. 1948. "Prediction of Pressure Drop During Forced Circulation Boiling of Water." *Trans. ASME*, Vol. 70, p. 695.

Martini, W.R. 1972. "Developments in Stirling Engines." Paper presented at the ASME Winter Annual Meeting, New York, August.

Matthews, C.W. 1972. Presented at the American Power Conference, Chicago, Illinois, April.

McAdams, W.H. 1964. *Heat Transmission*, p. 219, New York: McGraw-Hill Book Company, Inc.

McAlister, Roy E., 1991. "Renewable Electricity and Hydrogen Using Solar Thermal Processes." *Proceedings--Project Hydrogen '91 World Conference*, Independence, Missouri: American Academy of Science.

McAlister, Roy E. 1991. "Direct Hydrogen Injection and Spark Ignition System for Internal Combustion Engines." *Proceedings--Project Hydrogen '91 World Conference*, Independence, Missouri: American Academy of Science.

McCaull, J. 1948. "Windmills." *Environment*, Vol. 15, n.1, pp. 6-17, January.

McHardy, J., and Bockris, J.O.'M., *J. Electrochem. Soc.* 1973.

Meijer, R.J. 1965. "Philips Stirling Engine Activities." SAE International Automotive Engineering Congress, No. 65004, Detroit, Michigan, January.

Meinel, A.B. and Meinel, M.P. 1972. "Briefings Before the Task Force on Energy of the U.S. House of Representatives." Vol. III, U.S. Government Printing Office, Washington, D.C.

Mcinel, A.B. and Meinel, M.P. 1971. "Is it Time for a New Look at Solar Energy." *Bull. of Atomic Scientists*, Vol. 27, p. 32.

Meinel, A.B. and Meinel, M.P. 1972. "Physics Looks at Solar Energy." *Physics Today*, Vol. 52, p. 44.

Menard, W.A., et al. 1976. "New Potentials for Conventional Aircraft when Powered by Hydrogen-Enriched Gasoline." Proceedings 1st World Hydrogen Energy Conference, 5C, 59-86, Miami Beach, Florida, March 1-3.

Michel, J.W. 1973. "Hydrogen and Exotic Fuels." Oak Ridge National Laboratory, Report ORNI-IM-4461, June.

Miller, P.M. 1973. *Scientific American*, **228**, No. 2,78.

Mirza, M. Monirul Qader, Subrata, P. 1991. "Global Warming Sea Level Rise and Its Impact on Mangrove Ecosystem in Bangladesh." *Proceedings--Project Hydrogen '91 World Conference*, Independence, Missouri: American Academy of Science.

Moore, R.W., Fowle, A.A., Bailey, B.M., Ruccia, F.E. and Reid, R.C. 1960. "Gas-PressurizedTransfer of Liquid Hydrogen." *Advances in Cryogenic Engineering*, Vol. 5, p. 450, New York: Plenum Press.

Morgan, N.E. and Morath, W.D. 1965. "Development of a Hydrogen-Oxygen Internal Combustion Engine Space Power System." NASA CR-255, July.

Mucklow, G.F. 1927. "The Effect of Reduced Intake-Air Pressure and of Hydrdogen on the Performance of a Solid Injection Oil Engine." *R.A.S. Journal*, volume 31, No. 193, p. 17-59, January.

Murray, R.G., and Shoeppel, R.J. 1970. "A Reliable Solution to the Environmental Problem: The Hydrogen Engine." Proceedings SAE/AIAA/ASME Rel. & Maint. Conference, Detroit, Michigan, July 20-22.

Murray, R.G., Schoeppel, R.J. and Gray, C.I. 1972. "Hydrogen Engine in Perspective." Paper presented at the 7th Intersociety Energy Conversion Engineering Conference, San Diego, September.

Namba, I.K. 1962. "Development of Regenerative Compressor for Helium Circulation." Oak Ridge National Laboratory ORNL-TM-218, July 20.

Nema, S., Gontia, N., Khare, S., Nema, S.K. 1991. "Feasibility Studies of Photoassisted Hydrogen Generation on Membrane Supported Chlorophyll-Platinum." *Proceedings--Project Hydrogen '91 World Conference*, Independence, Missouri: American Academy of Science.

Nielsen, L. Henrik, Schleisner, Lotte, 1991. "Hydrogen as an Energy Carrier with Focus on Electricity Storage." *Proceedings--Project Hydrogen '91 World Conference*, Independence, Missouri: American Academy of Science.

Oemichen, M. 1942. Paper No. 68, *Verein Duetsche Ingenieur*.

Perkins W.E. and Frainier, R.J. 1960. "Practical Storage and Distribution of Liquid Hydrogen and Helium." *Advances in Cryogenic Engineering*, Vol. 5, p. 69, New York: Plenum Press.

Perris Smogless Automobile Association. 1971. "An Answer to the Air Pollution Problem...The Hydrogen and Oxygen Fueling Systems for Standard Internal Combustion Engines." First Annual Report, Perris, California.

Plass, G. 1972. *The Electrochemistry of Cleaner Environments*, Ed. J.O'M. Bockris, New York: Plenum.

Plass, G. 1956. *Tellus*, **8**, 140.

Putnam, P.C. 1948. *Power from the Winds*, Van Nostrand.

Ralph, E.L. 1972. "Large Scale Solar Electric Power Generation." *Solar Energy*, Vol. 14, p. 11.

Randall, R. and Sullivan, A. 1965. "A New Vacuum-Insulated Cryogenic Coupling." *Advances in Cryogenic Engineering*, Vol. 10, p. 451, New York: Plenum Press.

Reese, R.M. and Carmichael, A.D. 1971. "A Proposed Hydrogen-Oxygen Fueled Steam Cycle for the Propulsion of Dejep Submersibles." Presented at the 6th IECEC, Boston, Mass., August 3-5.

Regis, Ginalyn P. 1991. "The Energy Crisis: Fuel Cell Technology Viewpoint." *Proceedings—Project Hydrogen '91 World Conference*, Independence, Missouri: American Academy of Science.

Reilly, J.J. and Wiswall, R.H. 1968. "The Reaction of Hydrogen with Alloys of Magnesium and Nickel and the Formation of Mg_2NiH_4." *Inorganic Chemistry*, p. 2254-2256, #11, November.

Reilly, J.J. and Wiswall, R.H. 1973. "The Formation and Properties of Iron-Titanium Hydride." BNL-17711, January.

Reilly, J.J., Wiswall, R.H. and Aronson, S. 1968. *Metal Hydrides*, Brookhaven National Laboratory, Upton, New York, Report BNE-50149 (S-71), p. 34-39.

Reilly, J.J., Wiswall, R.H. and Hoffman, K.C. 1970. "Metal Hydrides as a Source of Hydrogen Fuel." Pres. Division of Fuel Chemistry, American Chemical Society, Chicago, September.

Reilly, J. and Wiswall, R.H. 1967. *Metal Hydrides*, Brookhaven National Laboratory, Upton, New York, Report BNL-50082 (S-70), p. 35-38.

Reilly, J.J. and Wiswall, R.H. 1974. "Formation and Properties of Iron Titanium Hydride," *Inorganic Chemistry*, Vol. 13, No. 1, pp. 218-222.

Reilly, J.J. and Johnson, J.R. 1974. "Titanium Alloy Hydrides: Their Properties and Applications." Brookhaven National Laboratories.

Reilly, J. and Wiswall, R.H. 1967. "The Reaction of Hydrogen with Alloys of Magnesium and Copper," *Inorgan. Chem.* 6, 2220.

Reilly, J.J., Hoffman, K.C., Strickland, G. and Wiswall, R.H. 1974. "Iron Titanium Hydride as a Source of Hydrogen Fuel for Stationary and Automotive Applications." Report BNL-18651, Brookhaven National Laboratory, April.

Reutschi, P. 1972. "Hydrogen Overvoltage." *Encyclopedia of Electrochemistry*, Robert E. Krieger Publishing Company, Huntington, New York.

Reutschi, P. 1972. "Oxygen Overvoltage." *Encyclopedia of Electrochemistry*, Robert E. Krieger Publishing Company, Huntington, New York.

Rex, R.W. and Howell, D.J. 1973. "Assessment of Geothermal Resources." in *Geothermal Energy*, Chap. 3, P. Kruger and C. Otte, Editors: Stanford University Press.

Reynolds, R.A. and Slager, W.L. 1972. "Transportation and Storage of Hydrogen for Eco-Energy." GE72TMP-54, General Electric Company, December.

Ricardo, H.R. 1953. *The High-Speed Internal Combustion Engine* (4th edition), London: Blackie & Son, p. 35.

Ricardo, H.F. 1923-24. Report of the Empire Motor Fuels Committee, *Proc. Inst. Automobile Engineers*, **18**.

Ricardo, H.F. 1923-24. "Further Note on Fuel Research." *Report of the Empire Motor Fuels Committee, Proc. IAE*, Vol. 18.

Richards, R.J., Steward, W.G. and Jacobs, R.B. 1960. "Transfer of Liquid Hydrogen Through Uninsulated Lines." *Advances in Cryogenic Engineering*, Vol. 5, p. 103, New York: Plenum Press.

Robinson, S.L. 1978. Sandia Laboratories, Report SAND77-8293, January.

Romero, J.B., Smith, D.W. and Dod, R.E. 1966. "Thermal Analysis and Optimization of Cryogenic Tanks for Lunar Storage Dewars." *Advances in Cryogenic Engineering*, Vol. 11, p. 231, New York: Plenum Press.

Rosen, B., Dayan, V.H., Proffit, R.L. 1970. "Hydrogen Leak and Fire Detection," NASA SP5092.

Rousseau, P.E. 1966. World Power Conference, Tokyo Section of Meeting, October 16-20, p. 912.

Ruckman, J.H., Billings, R.E., Woolley, R.L., Campbell, B.C., Hadden, L.D., Anderson, V.R. 1978. "Progress Report on Hydrogen Production and Utilization for Community and Vehicular Power." Billings Tech Library 78006, American Academy of Science, Independence, Missouri.

Rudolph, P.F.H., American Lurgi Corporation, Publ. 24072, 1974.

Russell, C.R., *Elements of Energy Conversion*, London, England: Pergamon Press, (1967) p. 378-388.

Salzano, F.J. (Editor) 1974. "Hydrogen Storage and Production in Utility Systems." First Annual Report, BNL-19249, July.

Sapru, K., 1991. "An Elementary Approach to Designing Metal-Hydrides for Practical Applications." *Proceedings--Project Hydrogen '91 World Conference*, Independence, Missouri: American Academy of Science.

Savage, R.L., et al, Editors, 1973. "A Hydrogen Energy Carrier." Vol. II, Systems Analysis, NASA-ASEE, NASA Grant NGT 44-005-114.

Schoeppel, R.J. 1972. *Chemtech*, p. 476, August.

Schoeppel, R.J. 1970. Paper presented at the Gas Symposium of the Society of Petroleum Engineers of AIME, Omaha, Nebraska, May.

Schoeppel, R.J. 1971. "Design Criteria for Hydrogen-Burning Engines." School of Mechanical and Aerospace Eng., Oklahoma State University, October.

Schoeppel, R.J., and Sadiq, S. 1970. Paper presented at the Frontiers of Power Technology Conference, Stillwater, Oklahoma, October.

Schuck, O.H., "The Future of Transportation." *AIAA Student Journal*, April.

Scott, R.B., et al. 1964. *Technology and Uses of Liquid Hydrogen*, Pergamon Press.

Scott, R.B. 1959. *Cryogenic Engineering*, p. 253, Princeton, New Jersey: D. Van Nostrand Company, Inc.

Senoo, Y. 1954. "Influence of the Suction Nozzle on the Characteristics of a Peripheral Pump and an Effective Method of Their Removal." Report of Research Institute for Applied Mechanics, Kyushu University, Vol. 111, No. 11, pp. 129-142.

Senoo, Yasutoshi 1948. "Theoretical Research on Friction Pump." Reports of Research Institute for Fluid Engineering, Kyushu University, Vol. 5, No. 1, pp. 23-38.

Senoo, Y. 1956. "A Comparison of Regenerative-Pump Theories Supported by New Performance Data." *Trans ASME*, Vol. 78, pp. 1091-1102, July (Paper 55-SA 44).

Senoo, Yasutoshi 1954. "Researches on the Peripheral Pump." Report of Research Institute for Applied Mechanics, Kyushu University, Vol. 111 No. 10, pp. 53-113, July.

Shaw, R.C. and Kobayashi, A.S. 1972. *The Surface Crack*, American Institute of Mechanical Engineers, New York, p. 79.

Sheean, T.V., and Steinberg, M. 1973. *Comparison of Different Types of Fluidised Bed Operations for Hydrogenation and Liquefaction of Coal*, Brookhaven National Laboratory, Upton, New York, October.

Shimosaka, Minoru 1960. "Research on the Characteristics of Regenerative Pump." *Bulletin of the Japan Society of Mechanical Engineers*, Vol. 3, No. 10, pp. 191-199.

Shimosaka, Minoru and Yamazaki, Shinzo 1960. "Research on the Characteristics of Regenerative Pump." *Bulletin of the Japan Society of Mechanical Engineers*, Vol. 3, No. 10, pp. 185-190.

Shoeppel, R.J. 1972. "Prospects for Hydrogen Fueled Vehicles." Oric, Non-Fossil Chem. Fuels Sym., *Amer. Chem. Soc.*, Bostom, Mass., April 9-14.

Silverstein, A., and Hall, E.W. 1955. "Liquid Hydrogen as a Jet Fuel for High-Altitude Aircraft." NACA Lewis Flight Propulsion Laboratory Report RM E55C 28a, April 15.

Small, W.J., Fetterman, D.F., and Bonner, T.F.Jr. 1973. Paper presented at the Intersociety Conference on Transportation; Denver, Colorado, September 1973.

Sofrata, H., Al-Saedi, Y., Aba-Oud, H., 1991. "A Laboratory Size Solar Hydrogen Generator." *Proceedings--Project Hydrogen '91 World Conference*, Independence, Missouri: American Academy of Science.

Sorensen, H. 1972. "The Boston Reformed Fuel Car - A Low Polluting Gasoline Fuel System for Internal Combustion Engines," Proceedings 7th Intersociety Energy Conversion Engineering Conference, San Diego, California, September.

Stoy, S.T. 1960. "Cryogenic Insulation Development." *Advances in Cryogenic Engineering*, Vol. 5, p. 216, New York: Plenum Press.

Strickland, G., Reilly, J.J., and Wiswall, R.H. Jr. 1974. "An Engineering-Scale Energy Storage Reservoir of Iron Titanium Hydride." The Hydrogen Economy Miami Energy Conference, Miami, Florida, March 18-20.

Strickland, G. and Reilly, J.J. 1974. "Operating Manual for the PSE&G Hydrogen Reservoir Containing Iron Titanium Hydride." BNL-18725, February.

Strickland, G., Reilly, J.J. and Wiswall, R.H. 1974. "Proceedings of the Hydrogen Economy Miami Energy Conference." University of Miami, Miami Beach, Florida, March 1974, p. S4-10.

Strickland, G. and Reilly, J.J. 1974. "Operating Manual for the PSE&G Hydrogen Reservoir Containing Iron Titanium Hydride." BNL-50421, February.

Strickland, G., Milan, J., Rosso, M.J. 1977. BNL23130, Department Applied Science, Brookhaven National Laboratory, August.

Strobridge, T.R. 1974. "Cryogenic Refrigerators--An Updated Survey." National Bureau of Standards, (U.S.) Tech Note 655, June.

Stuart, A.K. 1972. Paper presented at the American Chem. Soc. Symposium on Non-Fossil Fuels, Boston, April.

Swain, M.R., and Adt, R.R. 1972. Paper presented at the 7th Intersociety Energy Conversion Engineering Conference, San Diego, California, September. Paper presented at Intersociety Conference of Energy Conversion, University of Pennsylvania, Aug. 1973.

Swan, D.H., Velev, O.A., Kakwan, I.J., Ferreira, A.C., Srinivasan, S., and Appleby, A.J. 1991. "The Proton Exchange Membrane Fuel Cell--A Strong Candidate as a Power Source for Electric Vehicles." *Proceedings--Project Hydrogen '91 World Conference*, Independence, Missouri: American Academy of Science.

Swarnkar, M., Gupta, M.K., Mitra, V., Nema, S.K. 1991. "Investigation on SPE-Electrode Composite System for Solid Polymer Electrolyte Fuel Cell." *Proceedings--Project Hydrogen '91 World Conference*, Independence, Missouri: American Academy of Science.

Tegstrom, O., 1991. "Hydrogen Project Running Europe 1991." *Proceedings--Project Hydrogen '91 World Conference*, Independence, Missouri: American Academy of Science.

Telkes, M. 1955. "Solar Heat Storage." *Solar Energy Research*, F. Daniels and J.A. Duffie, Editors, University of Wisconsin Press, p. 57.

The Oil and Gas Journal, 23 July 1973, p. 27.

Thomas, S.R., Jr. 1974. "Experience in Operating a Hydrogen Fueled Internal Combustion Engine." Paper presented at Spring Meeting, Central States Section/ The Combustion Institute, University of Wisconsin, March.

Thompson, D.B. 1973. *Industrial Week*, 26 November, p. 17.

Timmerhaus, K.D. (Editor) 1960-1973. *Advances in Cryogenic Engineering*, Vols. 1-18, New York: Plenum Press.

Tiwari, S.K., Gupta, M.K., Mitra, V., Nema, S.K. 1991. "Evaluation of IME-SPE Hybrid Membranes for Hydrogen Production." *Proceedings--Project Hydrogen '91 World Conference*, Independence, Missouri: American Academy of Science.

Triona, A.R. 1960. *Transactions of American Society of Metals*, Vol. 52, p. 54.

Tseung, A. and Bevan, H. 1973. *J. Electroanalytical Chem*.

Tsujikawa, Y., Onoda, T., Fujii, S. 1991. "Performance Optimization of Scramjet Engine with Active Cooling." *Proceedings--Project Hydrogen '91 World Conference*, Independence, Missouri: American Academy of Science.

Underwood, P. and Dieges, P. 1971. "Hydrogen and Oxygen Combustion for Pollution Free Operation of Existing Standard Automotive Engines." Proceedings, 1971 Intersociety Energy Conversion Engineering Conference, Boston, August.

Ung, A.Y. and Back, R.A. 1964. *Can. Jour. of Chem.*, 42, 753.

Van Vucht, J.H.N., et al. 1970. "Reversible Room Temperature Absorption of Large Quantities of Hydrogen by Intermetallic Compounds." Philips Research Reports 15, 133-40, Eindhoven, The Netherlands, Phillips Research Laboratories, April.

Vander Arend, P.C. 1960. "Large-Scale Production, Handling, and Storage of Liquid Hydrogen."*Advances in Cryogenic Engineering*, Vol. 5, p. 49, New York: Plenum Press.

Vaughn, D.J. 1973. "`Nafion', and Electrochemical Traffic Controller" *DuPont Innovation*, E.I. duPont de Nemours and Company, Wilmington, Delaware.

Vavra, M.H. 1960. "Aerothermodynamics and Flow in Turbomachines." New York: John Wiley.

Vermishev, K. Kh. 1972. "Some Problems in the Long Range Forecasting of Power Generation and Solar Energy Utilization." *Applied Solar Energy*, Vol. 6, p. 1.

Veziroglu, T.N. Editor, 1974. *The Hydrogen Economy Energy Miami Conference*, Miami Beach, Florida.

Veziroglu, T.N. Editor, 1976. *First World Hydrogen Energy Conference*, Miami Beach, Florida.

Wall, T. and Feeney, R. 1975. "An Engineering Study of the Storage of Hydrogen on a Bus." ME 44 Report, Mechanical Engineering Department, Stevens Institute of Technology, May. (R.F. McAlevy, Advisor).

Walter, R.J. and Chandler, W.T. 1969. "Effects of High Pressure Hydrogen on Metals at Ambient Temperature." Report No. N70-18637, Rocketdyne, Canoga Park, California, February 28.

Watson, H.C., Milkins, E.E., and Deslandres, J.V. 1974. "Efficiency and Emissions of a Hydrogen and Methane Fuelled Spark Engine." Report T 8 74 Version 2, Mechanical Engineering, University of Melbourne, Australia (presented Fisita Conference, Paris, May.)

Weil, K.H. 1972. "The Hydrogen I.C.Engine-Its Origins and Future in the Emerging Energy Transportation-Environmental System." 1972 Intersociety Energy Conversion Engineering Conference, Paper No. 729212, San Diego, California.

Weinberg, A.M. 1973. "Long-Range Approaches for Resolving the Energy Crisis." *Mechanical Engineering*, Vol. 95, n.6, June.

Weinburg, A.M. and Hammond, R.P. 1970. *Am. Sci.*, July/August, p. 412.

Weinig, Frederich S. 1955. "Analysis of Traction Pumps." Wright Air Development Center, WADC Technical Report 54-554 (AD 67 339), June.

Weiss, S. 1973. "The Use of Hydrogen for Aircraft Propulsion in View of the Fuel Crisis" Technical Paper (NASA TMX-68242), paper presented and NASA Research and Technology Advisory Committee on Aeronautical Operating Systems, Ames Research Centre, Moffett Field, California, March.

Wilcox, D.E., Smith, C.I., Totter, H.C. and Hallett, N.C. 1968. Paper presented at the ASME Annual Aviation and Space Conference, June.

Williams, L.O. 1972. *Astronautics and Aeronautics*, p. 42, February.

Winsche, W., Wiswall, R.H., Reilly, J., Sheehan, T. and Winsche, W. 1968. "Metal Hydride Energy Storage Systems." Intersociety Energy Conversion Conference, Boulder, Colorado, IEEE Publ. 68C-21-Energy, p. 981-985.

Winston, Roland, Cooke, David, O'Gallagher, Joseph J. 1991. "Brighter than the Sun: New Solar Concentrator Technology." *Proceedings--Project Hydrogen '91 World Conference*, Independence, Missouri: American Academy of Science.

Wislecenus, G.F. 1965. "Fluid Dynamics of Turbomachinery." Dover.

Wiswall, R.H., Jr. and Reilly, J.J. 1972. "Metal Hydrides for Energy Storage," Report BNL-16889, Brookhaven National Laboratory, September.

Witcofski, R.D. 1972. "Hydrogen Fueled Hypersonic Transports." Presented at the American Chemical Society Annual Meeting, Symposium on Non-Fossil Chemical Fuels, Boston, Massachusetts, April 13.

Woerner, R.C. 1968. "A Study of Liquid Cylinder Filling Techniques." *Advances in Cryogenic Engineering*, Vol. 13, p. 237, New York: Plenum Press.

Wooley, R.L. et al. 1977. "Refueling Hydrogen Transit Fleets Part B - Data." Fourth International Symposium on Automotive Propulsion Systems sponsored by the NATO Committee on the Challenges of Modern Society, Washington, D.C.

Woolley, R.L., and Simons, H.M.1976. "Hydrogen Storage in Vehicles-- An Operational Comparison of Alternative Prototypes." SAE paper 760570, SAE Fuels and Lubricants Conference, St. Louis, June.

Woolley, R.L., and Henriksen, D.L. 1976. "Water Induction in Hydrogen-Powered IC Engines." First World Hydrogen Energy Conference, Vol. III, 6c-1, Miami Beach, Florida, March.

Woolley, Ronald L. and Hendriksen, D.L. 1976. "Water Induction in Hydrogen Powered IC Engines." Proceedings 1st World Hydrogen Energy Conference, 6C, 3-24, Miami Beach, Florida, March 1-3.

Woolley, R.L. 1976. "Performance of a Hydrogen-Powered Transit Vehicle," Billings Tech Library 76010, American Academy of Science, Independence, Missouri.

Woolley and Germane, "Dynamic Tests of Hydrogen-Powered IC Engines, "First World Hydrogen Energy Conference, Vol. III, 6c-59, Miami Beach, Florida, March 1976.

Wurm, J. and Pastris, R.F. 1974. "The Transmission of Gaseous Hydrogen." SPE 4526.

INDEX

A

absorb 74, 106
abundant 3, 5, 104
acid 76, 89, 95, 108, 110, 111, 132
acid rain 5, 106
acidic 110, 111
actuate 47, 102
adsorption 72, 114
advantage 6, 16, 46, 52, 54, 56, 96, 101, 112, 123, 124, 130, 138, 140
aerospace 47, 58, 61, 64, 84
affordable 76, 133
air 2, 4, 8, 11, 13, 16, 18, 20, 23, 27, 31, 32, 33, 34, 35, 36, 37, 38, 40, 41, 44, 45, 46, 49, 60, 62, 68, 69, 81, 89, 94, 104, 118, 119, 121, 123, 126, 133, 136
alcohol 2
alloy 66, 67, 68, 69, 70, 74, 75, 86, 91, 93, 96, 117
aluminum 70, 97, 100
American Academy of Science 132
ammonia 34, 82
analyzer 40, 41
anhydrous calcium sulfate 42
Ann Arbor, Michigan 81, 82
anode 107, 109, 111, 112, 113, 127, 133
Apollo Space Program 131
apparatus 11
apparent 64, 122, 130
appliance 4, 89, 118, 119, 120, 122, 123, 124
applicable 25, 68, 132
application 25, 38, 48, 53, 56, 57, 64, 72, 75, 79, 100, 105, 107, 114, 116, 129, 132, 134
Argosy Bus 96
Armstrong Machine Works Inc. 41
asbestos 109
atmospheric pressure 11, 39, 110
auto-hydrider 67
automobile 6, 16, 21, 26, 38, 61, 62, 64, 68, 79, 84, 102, 130, 132, 133
   diesel 130
   electric 33, 34, 76, 89, 95
   fuel cell 129
   gasoline 27, 130
   hydrogen 4, 7
   hydrogen conversion system 18
   pollution-free 3, 16
   solar 6
automotive 37, 66, 67, 68, 70, 75, 86, 132

automotive hydride tank 67
aviation 58, 64
Avogadro's number 108

B

backfire 18, 21, 28, 61, 80, 92, 95, 99
backflash 15, 18, 21, 23, 25, 26, 27, 28, 30, 32, 36, 46, 47
   control techniques
      independent ignition source 25
      reduction of spark plug gap 23
      shielded ignition cables 24
      single cylinder engines 25
      spark plug wire configuration 23
   source of
      carbon buildup 27
      rusting spark plug 27
balloon 2
barbecue grill 124
Barker, Lynn 14
base 55, 106
based 7, 11, 49, 97, 111, 116, 117, 132
batteries 33, 76, 89, 133
battery 133
beautiful 5
Beech Aerospace Corporation 61, 84
Billings Corporation 132
Billings, Lewis 11
blue-green algae 114
boil-off 59, 85
brake thermal efficiency 35
Briggs and Stratton 16, 78, 83, 84, 88
Brigham Young University 12, 30, 34, 35, 79
Brookhaven National Laboratory 66, 68, 89
Buick Skylark 101
burn 2, 4, 5, 7, 16, 19, 21, 23, 26, 27, 33, 34, 49, 52, 60, 63, 71, 89, 104, 106, 118, 119, 120, 122, 124, 126, 130, 140
burner 118, 119, 120, 121, 122, 123
burner quench diameter 118
bus 89, 95, 96, 99, 100
by-product 2, 4, 7, 104, 106, 124, 133

C

Cadillac Seville 91, 94, 124
calamity 7
calcium sulfate 42
calibrate 41
California 38, 96, 99
Caltrans 95
capacity 54, 60, 61, 74, 75, 96, 134

capture 85, 124
carbon 106, 119, 120, 131
carbon dioxide 7, 16, 56, 105, 106, 120, 131
carbon monoxide 16, 33, 34, 56, 72, 81, 105, 106, 119, 120
carbon steel 93, 100
carburetor 10, 11, 12, 13, 15, 21, 37, 38, 39, 44, 78, 82, 88, 91, 99, 100, 102
career 10, 13, 52, 69
catalyst 27, 104, 105, 107, 109, 117, 119, 121, 123, 124, 127, 129
catalytic 27, 46, 62, 64, 85, 119, 121, 122
catalytic reformer 55
categories 33
category 68
cathode 107, 111, 127, 133
cell 5, 6, 70, 71, 72, 95, 101, 102, 108, 110, 111, 112, 113, 126, 127, 128, 129, 130, 131, 132, 133, 134, 137, 140
Centigrade 32, 105
Cerritos, California 38
chamber 12, 13, 19, 21, 26, 27, 31, 34, 35, 36, 37, 44, 46, 48, 49, 50, 80, 81
Champion Spark Plug Company 27
charge 26, 27, 31, 45, 46, 49, 54, 60, 61, 69, 70, 71, 72, 85, 87, 93, 97, 113, 129, 130, 133
charge stratification 13, 46
Charles Kettering Foundation 84
chemical properties 4, 18, 21, 40, 52, 67, 72, 110, 126
Chevrolet 14, 15
Chevrolet Monte Carlo 61
chlorine 66
chromatograph 16, 80
circuit 109, 128
CO shift 105, 106
coal 5, 106, 107
coal gasification 3, 5, 105, 106
cold 60, 61, 68, 123
combustible 21, 36, 52, 60, 64
combustion 5, 7, 11, 13, 14, 16, 20, 21, 26, 27, 32, 34, 35, 36, 37, 40, 44, 45, 46, 47, 48, 49, 50, 52, 61, 69, 71, 78, 80, 81, 106, 119, 120, 121, 123, 126, 130
combustion chamber 31, 44
commercial 5, 53, 89, 105, 106, 107, 108, 109, 136, 137, 138, 139
communicate 63, 138
compact 53, 67, 76, 112, 133
compartment 82, 89
component 2, 3, 107
compress 26, 31, 110, 112, 130
compressed gas 53, 57

compressed gas cylinders 15
compression 46, 72
compression pressure 13
compression ratio 49, 78
compressor 60, 72, 110
computer 48, 49, 67, 68, 80, 102
computer control system 47, 49
concentration 16, 21, 33, 37, 82, 92, 113, 120, 121
condense 41, 113, 123
condensing 123
conserve 7
constant 13, 37, 63, 123
consume 2, 5, 31, 35, 47, 64, 101, 108, 110, 134
consumer 5
consumption 4, 7, 37
container 11, 41, 55, 58, 59, 60, 63, 68, 70, 71, 74, 96, 108
contract 47, 79, 80, 95, 96, 97, 99
conventional 4, 5, 6, 22, 23, 27, 44, 96, 118, 120, 123, 136
conversion 4, 5, 7, 21, 28, 30, 38, 41, 49, 50, 78, 79, 81, 89, 91, 96, 100, 102, 118, 122, 123, 126, 132
conversion kits 50
convert 4, 5, 6, 11, 15, 16, 30, 39, 44, 45, 55, 64, 67, 69, 78, 80, 81, 84, 88, 89, 90, 91, 96, 101, 102, 105, 106, 114, 119, 120, 123, 129, 130, 131, 134
cool 35, 41, 46, 54, 60, 82, 97, 100, 123
coolant 100
cooler 41
cooling 58, 71, 88, 93, 97, 117
copper 93, 100, 108, 109
corrosive 27
counterpart 45, 101, 130
creation 2
crisis 130
crowd 64
cryogenic 58, 59, 61, 63, 64
  storage container 58, 61, 84
current 16, 22, 53, 76, 108, 109, 110, 111, 128, 129, 133
custom 47, 62, 97
cylinder 11, 15, 22, 23, 25, 37, 42, 44, 45, 46, 47, 78, 89, 92, 95, 100, 102

D

Daimler-Benz 69
dangerous 7, 52, 60, 63, 68, 85, 119, 122
DCI 45, 46, 47, 49, 95, 96
de-oxo catalyst 72, 113
decision 2, 83, 84
decompose 5

Index 167

decrepitation 70, 93
degradation 72
demand 5, 8, 133, 134, 139
demonstrate 71, 78, 82, 89, 95, 114, 123, 136, 138
Department of Energy 137
destructive 7
detonation 2, 27
Detroit, Michigan 30
device 4, 10, 35, 42, 47, 49, 64, 67, 68, 108, 110, 113, 116, 117, 119, 129, 132
dewar 88, 102
DFVLR 102
diesel 33, 46, 47, 48, 49, 52, 130
diesel engine 49
diesel fuel 52
diesel fuel injector pump 47
direct cylinder injection 45, 47, 49, 95, 102
disperse 36, 37, 64
distributor 22
distributor cap 22
Dodge 88, 90
domestic 8
drain 42, 113, 123
drain pan 123
Drierite 42
driving 3, 62, 71, 86, 90, 91, 96, 131, 133
droplets 31, 32, 35, 36, 37, 44, 80, 84, 123
drought 7, 140
drying cylinder 42
dynamic 67
dynamometer 25, 40, 49, 82, 102

E

earth 5, 7, 104
education 138
effect 5, 7, 13, 46, 70, 111, 129, 131
efficiencies 4, 5, 129
efficiency 35, 46, 49, 56, 100, 108, 114, 123, 126, 129, 130, 140
efficient 5, 7, 70, 101, 104, 105, 106, 107, 108, 112, 114, 117, 120, 126, 129, 133
electric 34, 47, 48, 106, 108, 109, 111, 127, 128, 129, 130, 132, 133, 134
 lift truck 132
 motor 13
electrical 5, 79, 107, 108, 126, 127, 129, 133, 134
electricity 3, 5, 95, 107, 108, 110, 114, 126, 129, 133, 134
 grid 6, 134
electrocatalyst 111
electrode 109, 127

electrodes 108, 117
electrolysis 2, 3, 72, 107, 108, 109, 110, 112, 114, 129, 134
electrolyte 108, 109, 110, 111, 112, 113, 129, 132
electrolytic 112, 126, 128
electrolytically 109
electrolyzer 72, 108, 110, 111, 112, 113, 114, 129
 efficiency 108
 in Hydrogen Homestead 116
electromagnet 48
electromagnetic interference 22
electron 108, 111
electronic 47
electronically 47, 92, 102
electrons 127
electroosmotic effect 111
element 3
eliminate 3, 8, 16, 23, 25, 28, 32, 33, 34, 39, 46, 61, 80, 89, 95, 97, 119, 120, 121, 122, 124
emergency 63
Empire Motor Fuels Committee 13
encounter 31, 35, 69, 110, 124
energy 2, 3, 4, 5, 6, 7, 8, 19, 20, 21, 23, 31, 33, 35, 39, 48, 52, 53, 56, 58, 75, 78, 79, 89, 99, 104, 105, 106, 107, 108, 114, 118, 123, 124, 126, 130, 132, 133, 134, 137, 138, 140
 conservation of 7
 efficiency 105
 infrared 52
 sources of 3
 storage 3, 6
 utilization 104
energy crisis 130
energy efficiency 4
energy independence 134
Engelhard 95
engine 10, 11, 12, 13, 14, 15, 16, 18, 21, 22, 23, 24, 25, 26, 27, 28, 30, 32, 33, 34, 35, 36, 37, 40, 44, 45, 46, 47, 48, 49, 50, 52, 58, 61, 64, 67, 68, 69, 71, 78, 79, 80, 81, 82, 83, 84, 86, 87, 88, 89, 90, 91, 92, 93, 94, 95, 96, 97, 99, 100, 102, 126, 129, 130
engine efficiency 35
engineer 12, 18, 34, 37, 58, 67, 69, 79, 82
entrepreneurs 139
environment 7, 53, 56, 64, 104, 110, 111, 117, 121, 124, 126, 130
equipment 21, 58, 64, 80, 107, 108, 110
equivalence 32, 40, 41, 78, 88, 90
ethylene glycol 94
evaporate 61

exhaust 16, 21, 28, 33, 34, 40, 41, 42, 49, 54, 68, 71, 80, 86, 87, 92, 94, 95, 123, 124, 126, 130
expensive 41, 48, 58, 59, 67, 68, 87, 110, 123, 130, 132, 137, 139
experiment 10, 11, 12, 13, 14, 15, 19, 22, 25, 26, 37, 47, 66, 70, 80, 83, 95, 108, 123
explosion 2, 12, 21
external 33, 85, 97, 100, 110, 117, 128
extinguish 63
extrapolate 8
Eyring Science Center 80

F

fabricate 93, 133
Fahrenheit 58, 105
feedstock 5, 106
filter 26, 70, 93, 106
fire 2, 63, 64, 82, 86
fireplace log 124
flame 12, 23, 31, 35, 44, 46, 52, 62, 64, 67, 71, 118, 119, 120, 122, 123
  velocity 19
flame speed 16, 18
flames 52, 62, 63, 120
flammability 21
flammability range 20, 21
flammable 20, 21, 23, 121
flash-off 59, 60, 61
fleet 131
fleet vehicles 100
flywheel 33, 34
flywheel automobile 33
food 114
  production 8
  storage 8
Ford 16, 78, 79, 80, 84
Ford Falcon 84
Ford V-8 engine 79, 80
formation 4, 16, 32, 34, 37, 46, 49, 61, 80, 121, 122
fossil 136, 137, 140
fossil fuel 136, 137, 138, 140
founder 8
fuel 4, 7, 8, 10, 12, 13, 16, 18, 19, 20, 21, 22, 23, 26, 27, 28, 31, 37, 38, 39, 40, 41, 44, 45, 46, 47, 48, 49, 52, 53, 54, 55, 56, 58, 60, 64, 66, 69, 70, 71, 75, 80, 85, 88, 92, 95, 99, 100, 102, 104, 116, 119, 123, 124, 126, 128, 130, 131, 133, 136, 137, 138, 140
fuel cell 5, 70, 101, 126, 127, 128, 129, 130, 131, 132, 133, 134
  automobiles 6, 102

efficiency 129
phosphoric acid 95, 132
solid polymer electrolyte 132
fuel cell car 129
fuel cell lift truck 95
full power adjustment 39
function 32, 48, 54, 74, 90, 121, 129
funding 96, 132, 137
furnace 89, 123, 124
future 6, 8, 16, 49, 114, 134, 140

G

garage 12, 62, 63
gas 3, 4, 5, 6, 13, 15, 16, 34, 38, 39, 41, 42, 44, 47, 53, 54, 57, 58, 61, 66, 67, 68, 72, 79, 80, 89, 92, 104, 105, 106, 113, 114, 118, 119, 120, 122, 127, 130, 131, 134
gas chromatograph 80
gas ionization 23
gaseous 10, 55, 57, 58, 59, 60, 66, 69, 71, 87, 106
gasoline 10, 13, 16, 19, 21, 23, 26, 27, 37, 38, 44, 45, 46, 49, 52, 71, 89
  flame speed 16
General Motors 35
General Motors Proving Grounds 30, 33, 81
generate 3, 5, 34, 35, 36, 45, 47, 102, 106, 108, 110, 114, 122, 126, 128, 129, 133, 134
generating 55, 106, 110, 129, 134
generation 5, 28, 56, 95, 96, 100, 132
geothermal 107
Germany 69
global 7, 8, 28, 136, 140
global warming 7, 8, 136
government 137, 139
Grand Ville 68, 69, 87
greenhouse effect 7, 131

H

Hansen, John K. 8
heat 31, 35, 41, 58, 59, 60, 61, 68, 69, 71, 87, 88, 92, 93, 97, 100, 110, 118, 119, 123, 124, 130
heat exchanger 88, 94, 100, 118, 123
heating 67, 72, 117, 123, 124
Hindenburg 52, 138
Hindenburg syndrome 52
Holley carburetors 37
Holmes-Stretford unit 106
horn 62, 63
horsepower 35, 40
humidity 123, 124

Index   169

hydride   53, 54, 57, 66, 67, 68, 69, 70, 71, 72, 74, 75, 76, 85, 86, 87, 89, 90, 91, 92, 93, 94, 95, 96, 97, 100, 102, 113, 117, 129, 133, 137, 138. *See also* metal hydride
hydriding   67, 74, 86
hydrocarbon   8, 19, 20, 27, 55, 56, 79, 81, 105, 106, 107
   fuels   7, 18, 19, 20, 21, 27, 95, 106, 130
   reformation   53, 55, 104
hydrocarbons   20, 34, 56, 104
   unburned   16, 33, 34
hydrochloric acid   108
hydroelectric   3, 107
hydrogen
   appliances   116
      barbecue   124
      fireplace log   124
      furnace   123
      range   118
      water heater   123
   as energy storage   3, 7
   automobile   4, 7, 18
      power   44
   backfire   21
   backflash   21
   biological production   114
   bomb   52, 138
   bulk storage   117
   carburetor   10
   chemical properties   18
   combustion   2, 11, 119, 121
   commercial implementation   139
   conversion system   18
   diffusion   54
   efficiency   123
   electrolysis   3
   engine   13, 44
   flame   12, 19
   flame speed   16, 18
   flammability range   20, 21
   from coal   5
   from natural gas reformation   6
   from solar energy   6
   fuel   28
   gas   2, 7, 107
   ignition parameters   19
   implementation   136
   injector valves   48
   liquid   53, 64
      cryogenic storage containers   58
      refueling   59
      vacuum jacketed transport lines   60
   marketing   136
   metal hydride
      iron-titanium-manganese   71
      safety   71
   politics   99

   pollution free fuel   3, 8
   production   3
   purification   72
   safety   52
   solid   53
   space and aircraft applications   57
   storage   4, 53, 57, 104
      compressed gas   53
      iron-titanium-manganese hydride   70
      liquid   53
      liquid hydrides   53
      metal hydrides   53, 66
      micro-spheres   53
      on board hydrocarbon reformation   53
      solid hydrogen   53
   transportation of   5
   underground deposits   3
   unreacted   3, 104
   utilization   4, 7, 104
      practicality   136
   vehicles
      range   61, 85, 89, 90, 96
   world   3
hydrogen citicar   89
hydrogen compression   72
hydrogen energy   2, 13
hydrogen energy economy   3, 78
hydrogen engine   16, 18
   conversion   30
hydrogen enthusiasts   124
hydrogen equivalence ratio   32
hydrogen equivalents   142
Hydrogen From Coal - A Cost Estimation Guidebook   107
Hydrogen Homestead   91, 94, 116, 118, 122, 123, 124
   appliances
      barbecue grill   124
      furnace   123
      fireplace log   124
      lawn tractor   124
      natural gas range   118
      water heater   123
   electrolyzer   116
   metal hydride
      iron-titanium-manganese   117
   vehicles
      Cadillac Seville   124
hydrogen sulfide   5, 106
hydrogen technologies   136

I

idle mixture adjustment   39
ignite   12, 13, 19, 21, 22, 23, 26, 27, 31, 52, 64, 68, 71, 119
ignition   13, 19, 20, 22, 23, 24, 25, 27, 32, 46, 49, 64, 78, 119
ignition timing   49

Impco 38, 39
Impco carburetion system 61
Impco carburetor 37, 39, 88, 91, 100, 102
Impco Carburetor Corporation 37
implement 8, 24, 47, 49, 56, 99, 124, 129, 136, 137, 139, 140
impractical 68, 75
improvement 71, 92, 95, 109
incentive 137, 139
independence 134
Independence, Missouri 3, 100
induced 22, 23, 24, 25
induced spark ignition 22
induct 38, 92
induction 30, 31, 32, 33, 34, 35, 36, 37, 49, 61, 80, 90, 102
industry 69, 139
inexpensive 130
infrastructure 55, 130
Ingalls, California State Senator 96
injection 45, 47, 48, 80, 81, 84, 88, 90, 91, 92, 95, 100, 102
injector 47, 48, 49, 92, 102
injector system 47
injector timing 49
insulate 7, 58, 59, 60, 63, 117
intake 11, 12, 13, 21, 23, 26, 27, 36, 37, 44, 45, 46, 49, 92, 119
Intel Corporation 68
internal combustion engines 126
invisible 52, 122
ion 111
ionization 108
ions 111, 127
iridium 112
iron 68, 102, 105, 117
iron oxide 27, 105
iron-titanium 68
iron-titanium hydride 93
iron-titanium hydride vessel 89
iron-titanium-manganese 71
iso-octane 84

J

Jacobsen Lawn Tractor 94
jargon 18

K

Kansas 3
Kelvin 58
Ken Garth Mazda 81
Kohler 89, 94

L

L-head engine 16
laminar flame speed 18, 118

LaserCel 1 102, 133
launch 78, 84, 136, 138
lawn mower 12, 14, 15, 78, 94, 124
lawn mower engine 13, 15
　carburetor 10
　charge stratification 13
　conversion to hydrogen 10
LBL-type throttle 15
lead 8, 34, 76, 89
lead dioxide 117
liberate 4, 69, 108, 123, 126, 127
liquefied 60
liquid 10, 31, 37, 42, 44, 54, 55, 57, 58, 59, 60, 61, 62, 63, 64, 84, 85, 86, 87, 88, 102, 106, 108, 110, 113
liquid hydride 53, 54
liquid hydrogen 57, 58, 59, 60, 61, 62, 63, 64, 84, 85, 88
load 25, 40, 127, 128, 129
location 11
Los Alamos Airport Shuttle 101
Los Alamos National Laboratory 101
Los Angeles Times 35
loud 21, 28
Lucke, C.E. 13

M

magnesium 75
magnitude 7, 18, 19, 140
maintenance 58, 62, 64, 110, 113
malfunction 64, 86, 99
manganese 70, 71, 76, 93, 100, 102, 117
manifold 15, 23, 27, 36, 37, 44, 68, 92
manufacturer 47
map 84, 102
mass 27, 90, 91, 96, 97
mass transit vehicles 95
Max Mitchell 2
maximum 40, 59, 87, 128
Mazda RX-2 81
mechanical 12, 48, 79
mechanically 47
membrane 110, 111, 112, 113, 127
　solid polymer electrolyte 110
MERADCOM 132
Mercedes 69
metal hydride 53, 57, 66, 67, 68, 69, 70, 71, 72, 74, 75, 76, 85, 86, 89, 91, 92, 94, 95, 96, 97, 113, 117, 129, 133, 136, 137, 138
　alloy 67
　iron-titanium-manganese 70, 71, 96, 100, 102, 117
　mischmetal 68
　prototype 89
　R2D2 117
　safety 71

Index 171

weight percent 74
methane 7, 19, 20, 21, 104, 106, 118, 130
methanol 34, 55, 56
method 5, 16, 23, 24, 25, 28, 30, 34, 40, 46, 49, 53, 57, 58, 59, 61, 66, 69, 71, 72, 76, 100, 102, 110, 114, 119, 134, 139
micro-expansion measurement transducer 117
microcomputer 48, 68
microprocessor 68
microscopic 7, 35
microspheres 53
Mills
 California State Senator 96
Missouri 3, 100
Mitchell, Max 2
Model A 15, 16, 78, 79, 80
 backflash 15
 carburetor 15
 compressed gas cylinders 15
 exhaust samples 16
 L-head engine 16
 nitric oxide 16
molecular sieve 72, 114
molecular size 118
molecule 105, 108, 111
monitor 33, 34, 42, 48, 49, 101
Monte Carlo 61, 62, 66, 84, 85
moon 131
motor
 electric 13
movement 130

N

natural gas 6, 34, 38, 39, 61, 104, 105, 118, 119, 120, 122, 130, 131, 134
natural gas reformation 3, 104, 106, 134
Naval Ordinance Testing Laboratory 80
negative 34, 107, 109, 127
nickel 104, 109
nitric oxide 16, 28, 32, 33, 34, 36, 37, 46, 49, 61, 79, 80, 82, 92, 120, 121, 122, 123
nitrogen 16, 32, 33, 60, 120
 liquid 60
nuclear energy 3

O

observe 35, 72, 93
oil 48, 81, 84, 140
oil injection 81
oil spills 136
Onan generator 88

operate 16, 18, 25, 39, 45, 47, 54, 61, 62, 78, 79, 82, 85, 88, 89, 91, 92, 93, 95, 99, 100, 108, 110, 116, 118, 119, 120, 126, 129, 130
operation 7, 11, 27, 38, 46, 68, 89, 91, 92, 95, 99, 100, 106, 110, 111, 118, 122
optoisolators 49
order 11, 12, 18, 19, 24, 36, 37, 40, 63, 67
Orem, Utah 90
orifice 37, 47, 48, 118
oxide 66
oxygen 2, 3, 4, 16, 32, 40, 41, 42, 45, 49, 60, 72, 105, 107, 109, 110, 112, 113, 114, 120, 122, 126, 127, 129
 liquid 60
oxygen analyzer 40
oxygen overvoltage 112

P

panic 82
Paris, France 102
passenger 52, 89, 96, 97
patent 70, 72, 139
pattern 7, 140
peak 16, 23, 32, 35, 46, 129, 133, 134
peak voltage 23
perfluorosulfonic acid membrane 110
perform 12, 15, 48, 55, 56, 62, 82, 86, 100, 121, 137
performance 7, 22, 47, 49, 72, 84, 90, 91, 102, 112, 136, 140
perpetual motion 3
Peterson, Chuck 81
Peterson Motors 81
Peugeot 49, 102
Peugeot Sedan 102
phase 31, 35, 47, 100, 136, 137, 140
phenomenon 4, 13, 21, 25, 32, 35, 44, 45, 48, 53, 71, 74, 112
phosphoric 132, 133
photovoltaic solar collectors 114
pipe 5, 18, 19, 21, 60, 126, 134
pipeline 5
pistons 72
plastic 11, 63, 110
plate 92, 107, 118
plateau 74
platinum 112
pollutant 28, 33, 80
pollute 5
pollution 2, 3, 7, 8, 33, 34, 95, 99, 101, 106, 114, 130, 134
 air 2, 8, 33, 37, 126, 136
 nitric oxide 32, 79

polymer 110, 111, 113, 127, 129, 132, 133
Pontiac 86
Pontiac Grand Ville 68, 69, 87
position sensor 47
positive 107, 109, 127, 138
Postal Jeep 99, 100
potassium hydroxide 108, 109
potential 44, 53, 55, 56, 58, 63, 68, 70, 79, 82, 95, 99, 108, 128, 138
powdered 66, 68
power 2, 5, 6, 15, 26, 39, 40, 44, 45, 46, 47, 48, 52, 54, 88, 90, 95, 96, 99, 106, 112, 114, 123, 126, 128, 129, 130, 131, 132, 133, 134
practical 28, 57, 59, 130, 136, 140
pre-ignition 13, 25, 27
press 23, 63
pressure 11, 13, 35, 38, 39, 46, 47, 48, 54, 58, 59, 60, 61, 62, 63, 69, 72, 74, 87, 97, 100, 110, 112, 129
pressure swing adsorption 72, 113
prevent 23, 25, 27, 60, 62, 84, 85, 109, 110, 118, 119
price guarantee 139
primary air 89, 118, 119, 120
primary air intake 119
principal 7
problem 7, 8, 16, 18, 21, 23, 24, 25, 27, 30, 36, 37, 45, 46, 47, 59, 60, 66, 68, 69, 70, 79, 82, 83, 84, 85, 86, 93, 99, 119, 120, 121, 124, 136, 137, 138, 140
problems 16
process 3, 5, 31, 55, 56, 58, 60, 61, 70, 72, 104, 105, 106, 107, 110
produce 3, 5, 6, 7, 22, 35, 52, 72, 81, 104, 106, 107, 108, 109, 110, 112, 114, 116, 117, 122, 130, 134
program 58, 80, 131, 137, 138
propane 34, 37, 38, 39, 89
properties 4, 18, 21, 40, 52, 55, 66, 67, 72, 110, 126
property 20, 66, 121
proposal 79, 96
propulsion 55, 78, 126, 129
protons 111, 127
prototype 15, 16, 47, 49, 61, 62, 66, 69, 78, 79, 82, 84, 85, 86, 87, 89, 91, 92, 93, 100, 102, 133, 137
Provo, Utah 64, 69, 81, 83, 89, 90, 95, 116
Provo-Orem Bus 89
public 8, 52, 90, 95, 137, 138
published 13, 84

pump 47, 48, 55, 61, 104
pumping 7, 46, 59

Q

question 14, 18, 52, 64, 136

R

radiator 97, 126, 130
Rampton, Governor Calvin 83
Rankine-cycle device 35
ratio 11, 16, 32, 37, 40, 44, 49, 78, 88, 90, 107
react 45, 54, 62, 66, 67, 69, 104, 105, 106, 113, 121, 127, 133
reactant 129
reaction 19, 31, 33, 67, 95, 105, 106, 122, 127, 128
reactivate 72
reactive 66, 74
reactor 85
recalibrate 82
recharge 60, 69, 70, 72, 87, 93, 97, 99, 113, 129, 130, 133
recharging 54, 69, 72, 97
recirculation 34, 80
reduce 8, 16, 23, 25, 28, 32, 35, 38, 45, 46, 69, 70, 87, 112, 131
reduction 4, 23, 45, 70, 80, 93, 106
reformation 3, 6, 55, 56, 104, 105, 106
refrigeration 113
regenerate 106
Reilly, J.J. 66
reliable 72, 132, 137
remove 10, 26, 34, 37, 42, 62, 72, 82, 89, 93, 106, 110, 113, 114
renewable energy 7, 107, 114
reproduce 22
residential 124
retrofit 49, 102, 130
reverse 7, 8, 35, 129
reversible 67
Ricardo, H.R. 13
Ridge, Dr. Robert 48
Riverside Bus 95
Riverside, California 95, 96, 99, 100
rotor 22
rotten 5

S

sabotage 99
safe 52, 53, 55, 62, 64, 66, 67, 69, 71, 76, 87, 104, 136
safety 52, 53, 63, 64, 71, 133, 138
salt 66, 67, 81
Salt Lake City, Utah 81, 83
sand 99
saturate 74, 113

science 2, 3, 11, 12, 13, 14, 15, 16, 78, 79, 80, 83, 133, 136, 138
science educators 138
science fair project
  lawn mower engine 10
scientific solutions 136
scientist 2, 3, 4, 7, 108, 136, 137, 138, 139
scouring pad 121
secondary air 119
sensitive 41
Seville 91, 93, 94, 124
sieve 72, 114
signal 15, 63
silica 99
sintered metal filter 70
slush 53
sodium 66
solar energy 3, 6, 107, 130, 134
  photovoltaic collectors 114
solenoid 47, 48, 86
solid 53, 110, 111, 112, 113, 127, 129, 132, 133
solid hydrogen 53
solid polymer electrolyte 110
  perfluorosulfonic acid membrane 110
source 3, 4, 5, 7, 8, 23, 25, 27, 32, 46, 56, 64, 71, 104, 107, 114, 129, 132, 133, 134, 137
space shuttle 58, 132
spark gap 23
spark ignition 78
spark plug 19, 22, 23, 25, 27, 31, 49
  wire configuration 23
SPE 110, 112, 113, 114, 116, 127, 132
specifications 47, 87
spring 39, 82
stainless 27, 68, 70, 117, 121, 122, 123, 124
stainless steel 68, 86, 91, 93, 117, 121
Star Wars 117
stationary 7, 57
steam 5, 31, 35, 104, 105, 106, 123, 124
steam engine 35
steel 27, 68, 70, 86, 91, 93, 100, 117, 121, 122, 123, 124
storability 6
storage 7, 8, 53, 54, 55, 57, 58, 59, 60, 61, 64, 66, 67, 68, 69, 70, 71, 72, 74, 75, 84, 85, 86, 91, 92, 96, 113, 117, 129, 133, 134, 138
substantial 16, 28, 32, 47, 70, 93, 99, 100, 105, 106, 107, 110, 112, 137
suffering 8
sulfide 106
sulfonate 111
sulfur 5, 106
sulfur dioxide 106

sulfuric acid 111
sun 6, 114
sunrise 62
Superbeetle 30, 81, 82
surface-to-volume ratio 16
synthesize 4
synthetically 3
system 7, 12, 13, 14, 15, 18, 19, 21, 22, 23, 24, 26, 28, 32, 36, 37, 38, 39, 42, 44, 46, 47, 48, 49, 53, 54, 60, 61, 64, 66, 67, 68, 69, 70, 71, 72, 78, 81, 85, 86, 87, 88, 89, 90, 91, 92, 93, 96, 97, 99, 100, 107, 110, 111, 113, 114, 116, 123, 129, 130, 133, 138, 140

T

tailpipe 21, 126
tank 52, 55, 59, 60, 61, 62, 63, 67, 69, 70, 71, 72, 74, 84, 85, 86, 87, 97, 100
tax break 139
teacher 2, 3
technological breakthrough 69
technology 8, 31, 37, 49, 54, 56, 57, 69, 70, 79, 81, 94, 95, 96, 100, 101, 106, 110, 112, 116, 132, 133, 136, 137, 138, 139
temperature 16, 19, 20, 27, 32, 35, 36, 46, 54, 58, 59, 60, 74, 75, 92, 93, 104, 117, 119, 120, 121, 122, 129
thermal 5, 27, 35, 107
thermodynamics 3
throttle 15, 46, 49, 82, 92
titanium 68, 71, 75, 76, 100, 102, 117
transit vehicles 89, 95. *See also* mass transit
transportation 8, 53, 95
trap 41, 42, 113
trend 7, 8
trigger 47
tube 11, 13, 68, 86, 87
turbine 5
turbo-charged 46
twin 102

U

U.S. Post Office 99
UCLA 34
underground 55, 134
uniform 37
uniformly 36
Union Carbide 88, 114
unit conversions 145
United Catalysts 113
unsolicited 79

Index 173

urban 5, 8, 30, 33, 81, 90, 96, 99, 134
Urban Vehicle Design Competition 30, 33, 81
Utah 12, 64, 69, 81, 83, 90, 95, 99, 116
Utah Academy of Science, Arts, and Letters 12
Utah State Mental Hospital 83
utilization 4, 5, 6, 7, 25, 39, 70, 105, 123, 129, 130, 140
utilize 3, 4, 5, 16, 30, 33, 35, 48, 49, 61, 71, 102, 105, 107, 114, 116, 117, 122, 126, 129, 130, 131, 134
utilizing 4, 33, 47, 54, 56, 61, 68, 69, 80, 88, 94, 107, 114, 121, 130, 136

V

vacuum 11, 58, 60, 61, 63
valve 11, 13, 15, 23, 26, 42, 46, 48, 49, 60, 62
vapor 31, 35, 41, 61, 62, 72, 113, 124, 133
vaporize 35, 58, 59, 60, 61, 63, 88, 123
vaporizer 61
variable 78, 88, 129
vehicle 7, 15, 16, 21, 23, 30, 33, 34, 49, 53, 54, 55, 57, 59, 61, 62, 63, 64, 66, 67, 68, 69, 70, 71, 72, 76, 78, 79, 81, 82, 85, 86, 87, 88, 89, 91, 95, 96, 97, 99, 100, 101, 102, 104, 106, 126, 129, 130, 131, 132, 133, 136
vent 61, 85
Verne, Jules 139
vessel 54, 58, 59, 60, 61, 62, 63, 64, 68, 71, 72, 74, 75, 84, 85, 86, 89, 90, 91, 93, 94, 95, 97, 100, 102, 113, 117, 118, 129, 133, 138
vested interests 99, 137
viscosity 118
Volkswagen 14, 30, 33, 34, 82
  Superbeetle 30, 34, 81
volt 128
voltage 22, 23, 25, 49, 108, 109, 110, 112, 128, 129
volume 11, 16, 38, 39, 40, 44, 47, 54, 57, 118
volumetric 46
voluminous 44, 45, 54, 57, 59, 61

W

Wankel engine 81
war 140
Washington, D.C. 89
water 2, 3, 4, 11, 31, 32, 33, 34, 35, 36, 37, 38, 41, 42, 62, 63, 67, 71, 80, 81, 84, 88, 89, 90, 92, 93, 97, 104, 105, 107, 108, 109, 110, 111, 113, 114, 116, 123, 126, 128, 129, 133
water heater 123
water induction 30, 33, 35, 36, 49, 61, 90
water injection 80, 84, 90, 91, 100, 102
water injection system 81, 88
water jacket 97, 100
water trap 41, 113
water vapor 72, 124
weight 57, 58, 70, 74, 75, 76, 89, 91, 97, 106, 111, 129
weight percent 74, 91
weld 11, 71, 86, 100
wind 3, 134
wind energy 107
Winnebago 8, 88, 89, 90
Wiswall 66
world's first hydrogen car 79